Energy Storage

edited by
Md. Rafiqul Islam Sheikh

Energy Storage
Edited by Md. Rafi qul Islam Sheikh

This is a reprint of the book published in 2010 by Sciyo

Republished by InTech
Janeza Trdine 9, 51000 Rijeka, Croatia

Publishing Process Manager Jelena Marusic

Technical Editor Goran Bajac

Cover Designer Martina Sirotic

Additional hard copies can be obtained from orders@intechopen.com

Energy Storage, Edited by Md. Rafi qul Islam Sheikh
p. cm.
ISBN 978-953-307-119-0

For EU product safety concerns:
IN TECH d.o.o., Prolaz Marije Krucifikse Kozulić 3, 51000 Rijeka, Croatia,
info@intechopen.com or visit our website at intechopen.com.

Contents

Preface

Electricity is more versatile in use because it is a highly ordered form of energy that can be converted efficiently into other forms. However, the disadvantage of electricity is that it cannot be easily stored on a large scale. One of the distinctive characteristics of the electric power sector is that the amount of electricity that can be generated is relatively fixed over short periods of time, although demand for electricity fluctuates throughout the day. Almost all electrical energy used today is consumed as it is generated. This poses no hardship in conventional power plants, where the fuel consumption is varied with the load requirements. However, the photovoltaic and wind, being intermittent sources of power, cannot meet the load demand all of the time, 24 hours a day and 365 days of the year. Wherever intermittent power sources reach high levels of grid penetration, energy storage becomes one option to provide reliable energy supplies.

Developing technology to store electrical energy would represent a major breakthrough in electricity distribution as the stored energy can be available to meet demand whenever needed. Helping to try and meet this goal, electricity storage devices can manage the amount of power required to supply customers at times when need is greatest. These devices can help to make renewable energy more smooth and reliable, though the power output cannot be controlled by the grid operators. They can also balance micro grids to achieve a good match between generation and load demand. Storage devices can provide frequency regulation to maintain the balance between the network's load and power generated, and they can achieve a more reliable power supply for high tech industrial facilities. Also, it can significantly improve the load availability, a key requirement for any power system. The energy storage, therefore, is a desired feature to incorporate with renewable power systems, particularly in stand alone power plants.

The purpose of this book is twofold. At first, for the interested researcher it shows the importance of different Energy Storage devices, but secondly, and more importantly, it forms a first attempt at dissemination of knowledge to the wider non-expert community who may wish to consider Energy Storage device for specific application. Thus this book will be helpful to provide an indication of the tools necessary for an assessment to be made Energy Storage device more powerful.

In Chapter 1, different topologies of electric machines which are used in Energy Storage systems have been described. Among the various Energy Storage systems- pumped hydro storage, compressed air energy storage (CAES) and flywheel energy storage system (FESS) have been discussed extensively. Also different machine topologies suitable for specific application have been presented.

The beginning of the twenty-first century is an exciting time for wind energy. With the changes in technology, policy, environmental concern and electricity industry structure which have occurred in recent years, the coming decade offers an unparalleled opportunity for wind energy to emerge as a viable mainstream electricity source and a key component of the world's environmentally sustainable development path. However, the lack of controllability over the wind and the type of generation system used cause problems to the electric system. Therefore, Chapter 2 presents a detailed model and a multi-level control of a DSTATCOM controller coupled with FESS to improve the integration of wind generators into the power system. The DSTATCOM/FESS device is presented with all of its components in detail. Moreover, the complete control for this device is suggested to control voltage, power factor and to minimize output fluctuations of wind farm.

In Chapter 3, different energy storage devices like superconductive inductor, super capacitor, battery have been discussed briefly. Finally, the principle of high-speed FESS, factors affecting the FESS efficiency and its control systems have been discussed extensively.

Chapter 4 covers the characteristics of a grid-connected photovoltaic system and the advantages to use it with battery energy storage system. The main idea that is presented in this work is that batteries can be used in a new and more convenient "distributed manner" as distributed passive maximum power point tracker (MPPT) devices, in alternative to more expensive active MPPTs.

In Chapter 5, detailed modeling and control strategies for superconducting magnetic energy storage (SMES) system have been discussed for improvement of load frequency control in multi-area power system. The advantages and disadvantages of SMES system have also been described. Finally, it is presented and evaluated that SMES is more effective to enhance the stability of power system as it has both active and reactive power control abilities with high response speed. Thus can be a good tool for load frequency control application.

Chapter 6 discusses about the air pollutants that are released from different sources like coal, oil, motor vehicles, diesel engine exhaust, paper mills and natural gas-burning electric power plants etc. and presented the harmful effects of these air pollutants. An experimental study on nitrogen oxide (NO) removal in a pulse corona discharge reactor is presented to clarify the influence of the streamer-to-glow transition on NO removal, where the inductive energy storage pulsed power generator is used. The technique for improving the energy efficiency has also been discussed.

Finally, in Chapter 7, aquifer thermal energy storage (ATES) system has been discussed. It is presented that large scale thermal energy storage can be accomplished in the aquifer through the installation of an array of vertical boreholes. A numerical investigation and thermo hydraulic evaluation of two-well models of aquifer thermal energy storage system under continuous flow regime have also been presented.
At the end, the book gives numerous references for further reading.

Editor

Md. Rafiqul Islam Sheikh
Associate Professor
EEE Department
Rajshahi University of Engineering & Technology
Bangladeshl

1

Electric Machine Topologies in Energy Storage Systems

Juan de Santiago and Janaína Gonçalves de Oliveira
Division for electricity, Uppsala University
Sweden

1. Introduction

Energy storage systems based on pumped hydro storage, compressed air (CAES) and flywheels require electric machines working both as motors and generators. Each energy storage system has specific requirements leading to a variety of electric machine topologies.

Hydro power and CAES stations have several configurations; they may have a turbine-generator and an independent pump-motor group or a common turbine-motor/generator assembly, but in both cases the electric machines are coupled to turbines that are operated at constant speed and low electric frequency at steady state. Synchronous machines are the predominant technology for these applications.

Modern flywheel concepts based on a composite rotor driven by an electric machine started to be studied in the 1970s and 1980s. It is therefore a relatively new field of research based on the latest developments in strong light weight materials, new magnetic materials, magnetic bearings and power electronics. Despite the short history of the concept, there are already commercial applications and other potential applications have been identified such as space applications, Uninterruptible Power Supply (UPS), vehicles, grid quality enhancement, integration of renewable sources, etc. Flywheels are operated at high and variable speed and require specific machine topologies. Permanent magnet machines are preferred for vehicular flywheel applications (Acarnley et al., 1996), although inductance and reluctance topologies are applied for stationary flywheels.

Most common and promising types of machines use in energy storage systems discussed in this chapter are presented in Table 1.

	Type	Properties
Constant Speed	Synchronous	- Well established technology. - Unlimited power rate.
Variable Speed	Induction	- Robust and no iddle losses. - Lower efficiency than other topologies.
	Permanent Magnet	- Highest efficiency and power density. - Sensitive to temperature. - Higher material price.
	Reluctance	- Robust and no iddle losses. - Complex control.

Table 1. Electric machines used in Energy Storage applications.

The purpose of this chapter is to discuss newly research threads and specific aspects in energy storage applications. For a general overview of the synchronous machine and a much detailed discussion of synchronous motor and generators, consult (Rashid, 2007), (Laughton & Warne, 2003).

2. Constant speed operation machines

Due to the specific orography and water flow at the location, every hydro power station is unique and required tailor maid solutions in terms of water head and flow. Pumped hydro power plants may be equipped with an independent pump with a specific motor (generally an induction motor) or with reversible pump-turbine and a single motor/generator machine. Hydro turbines are directly coupled to the generator shaft without an intermediate gear box. The low speed operation of turbines forces a high number of poles in the electric machine in order to run at synchronous speed with the grid frequency.
Currently operated CAES are integrated in hybrid power plants. The air is compressed and stored in a reservoir when the electricity price is low, to be mixed with fuel and expanded in a conventional gas turbine at peak demand. Air pumps are operated at higher speeds than hydro pumps and therefore cylindrical rotor turbo machines are used in CAES power plants.
In both hydro power station and CAES, the turbine speed is constant at steady state operation. Synchronous machines are optimal for constant speed operation and dominate the high power station market.

2.1 Synchronous machines

In synchronous motor/generators, the rotor is wound and a DC current creates the rotor magnetic field. The rotor may be essentially described as an electromagnet. The magnetic field induced by a DC current is intrinsically invariant; the rotational movement makes the magnetic flux vary in time through the stator windings. The cross section of this kind of machines is shown in Fig. 1.

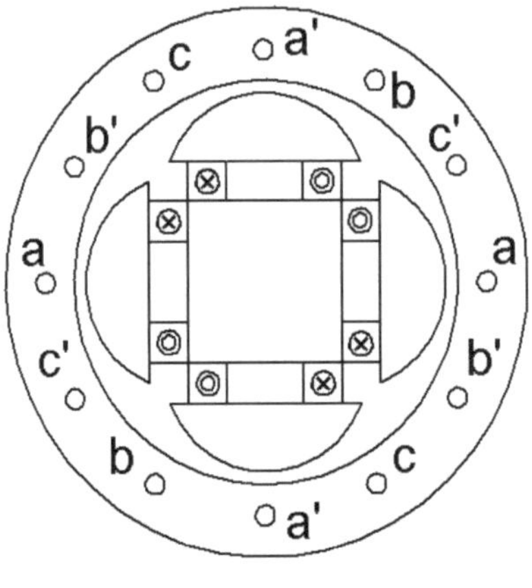

Fig. 1. Cross section of a three phase generator with a four salient pole rotor.

The mutual inductance between rotor and stator coils and pole saliency induces the electromotive force (e.m.f.) in the stator windings. For a three phase salient pole synchronous motor with negligible stator winding resistance, the electromagnetic power is expressed as (Laughton & Warne, 2003):

$$P = 3 \cdot \left[\frac{V \cdot E}{X_{sd}} \cdot \sin(\delta) + \frac{V^2}{2} \cdot \left(\frac{1}{X_{sq}} - \frac{1}{X_{sd}}\right) \cdot \sin(2 \cdot \delta) \right] \tag{1}$$

Where V is the input phase voltage, E is the e.m.f. induced by the rotor excitation flux or open circuit voltage, and δ is the power angle or angle between E and V; X_{sq} and X_{sd} are the synchronous reactances in the d axis and q axis. The equivalent one phase circuit of the synchronous generator may be represented as in Fig. 2.

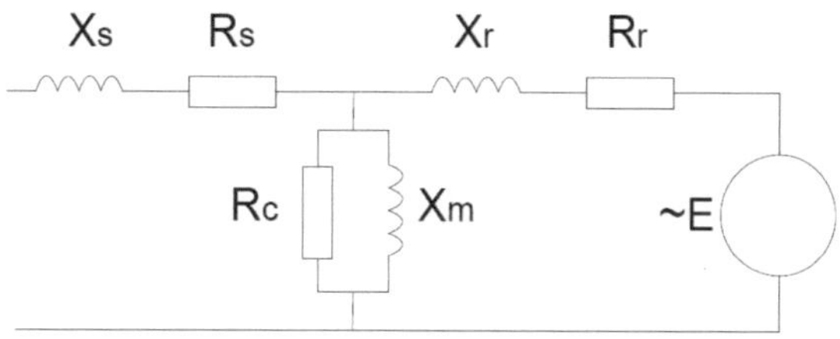

Fig. 2. One phase equivalent circuit of a synchronous generator

Even though synchronous generators are a mature technology and efficiencies up exceeding 98% have been reported, there are important research threads in this type of machines as described in following sections.

2.1.1 High voltage insulation systems

The stator is formed with a three phase winding. The armature windings in the stator are made of copper bars and packed as tight as possible to achieve a high filling factor (copper cross section/bar cross section). Due to the limited permeability of the laminated steel in the stator, the electric field induced in the stator bars depends on the vertical position of the slot. To equalize the voltage induced in each of the strands and eliminate circulating currents, they are usually transposed. Modern generators use the so called Roebel transposition. Every copper strand is insulated and strands are packed into bundles. High power rated generators have hollow copper tubes in the bundles for water or gas cooling. Insulation between copper bars is used to avoid short circuits but also to prevent corona effect. The insulation layers are made with different materials, traditionally based on mica. The insulation materials limit the generator voltage rates. There are several standard ratings (Changda et al., 1998). Even high power rated generators rarely exceed 25 kV so transformers to couple the grid voltage are required. As an example, generators at the Three Gorges dam are rated over 700 MVA at only 20 kV. These generator low voltage rates force high nominal currents that cause a significant amount of generator total losses.
A new technology proposes to wind the stator with high voltage, dielectric insulated cables, to withstand higher voltage ratings. This technology is known as Powerformer. High voltage operation increases overall efficiency and avoids the need of transformers. This technology is particularly interesting for energy storage systems with independent motor and generator machines in stationary systems. Motors and generators have different ratings and therefore different machine solutions may be adopted.

2.1.2 Multiphase systems

Increasing the voltage is not the only strategy in the windings design to improve the performance of generators. Multiple phase systems and more than one set of windings in the stator have been proposed and currently under development. There are several advantages in multiple phase systems from the generators point of view. With multiphase systems, the magnetic field distribution in the air gap is more homogeneous and the power is distributed into more phases, reducing the current in every phase. Note that the reduction in current per phase does not reduce the current density nor the Joule losses as the slots in the stator have to be divided into more phases. The improvement of lower currents per phase relies in the lower power ratings in inverters and lower short circuit current in case of fault.

Designers are usually restricted to the three phase system as generators have to match the standard electric grid three phase system. Nevertheless, there are several threads of investigation in this field.

The Powerformer, discussed in section 2.1.1, has two sets of independent windings in the stator at different voltage levels. The main windings deliver power to the grid at high voltage, higher than the ancillary services. To supply different plant equipment, the Powerformer may have a devoted set of windings generating at lower voltage rate (Touma-Holmberg & Srivastava, 2004).

The first application for two winding generator was developed as early as 1920's. In order to lower fault current in large generators and allow electrical segregation of bus sections in power stations, two identical layers of three-phase winding were proposed. Nowadays the same idea has been adopted to decrease the power rating of high power traction drives. Stators designed with a double star stator configurations require two power inverters but at half of the power rate.

The double winding configuration is also applied for inductance machines. In both cases the optimal angle between windings has been calculated in 30 electrical degrees (Fuchs & Rosenberg, 1974). Ground of both star windings are connected, resulting an equivalent circuit as shown in Fig. 3.

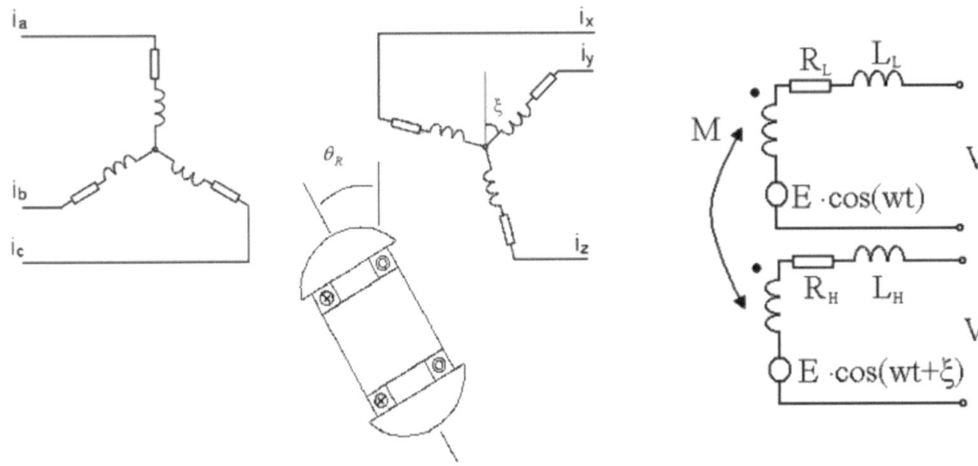

Fig. 3. Equivalent circuit and one phase equivalent circuit of a double star winding synchronous generator.

2.1.3 Excitation system

The excitation system provides a DC current into the field winding of the generator to produce the magnetic field in the rotor. This apparently simple device has been classified in 12 different types of excitation systems by the IEEE standards (Kim, 2002), (IEEE Std 421.1-2007). The complexity of the excitation system lies on the control and regulation techniques. The field current regulates the no load voltage and the reactive power delivered by the generator. Modern excitation systems tend to avoid graphite brushes; the slip rings are replaced by a multiphase set of windings and the power is transfer to the rotor through the magnetic fields induced in the exciter stator. The AC currents in the rotor are rectified in rotating rectifiers mounted in the shaft and create the DC field current in the generator. The magnetic field in the excitation system stator is produced by a controlled current, either from a synchronous generator or a transformer connected to the generator's terminals. An alternator-rectifier exciter scheme is presented in Fig. 4. The AC exciter current is rectified in passive rectifier bridges. Controlled thyristors mounted on the rotor have been proposed, but this technology is still not commercially developed as it significantly increases the generator costs.

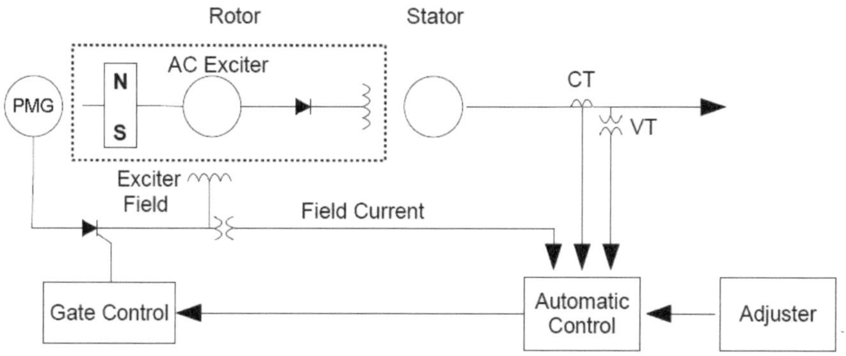

Fig. 4. Alternator-rectifier exciter employing rotating non-controlled rectifiers.

Standard excitation systems are based on a DC current that flows through a single phase field winding, but more complex configurations are also possible. Two phase excitation systems have been proposed to create a rotating magnetic field in the rotor. The magnetic field rotational speed that would see the stator windings is the addition of the mechanical rotational speed plus the magnetic field circulation around the rotor. This machine is called asynchronised synchronous generator. They are designed to operate up to a maximum slip of 20%. The speed regulation is particularly interesting in hydro generators with wide range of water head changes and gas turbines to operate them with a low inertia constant (Mamikoniants et al., 1999).

Multiple phase rotors may be also be designed to improve the magnetic field distribution in the airgap. The magnetic field distribution in simple or double phase excitation systems rely on the symmetry of the rotor and stator geometry. The excitation field current provides the magnetomotive force in the magnetic field circuit that flows through rotor and stator. An eccentricity or miss aliment in the rotor would create a non uniform magnetic reluctance and therefore unsymmetrical magnetic field distributions. The region where the magnetic field increases suffers saturation in the teeth steel which leads to harmonics in the e.m.f., higher hysterics losses and higher current in dumping bars. Eccentricity in the rotor leads also to unbalanced radial forces and wear (Lundin & Wolfbrandt, 2009). Multiple and independent field winding phases controlled by rotating thyristors may be a solution for rotor eccentricity.

3. Variable speed operation machines

Flywheel energy storage systems are base on the variation of rotational energy with rotational speed. Almost constant speed flywheels with synchronous generators, with a speed deviation of around 2% of the nominal speed, have been studied, but the high moment of inertia required make this configuration impractical (Carrillo et al., 2009). Therefore flywheels are designed to vary speed with a maximum nominal speed of about twice the minimum speed, and require variable speed machines. The speed range varies from applications, but generally nominal speeds are over the standard 50 or 60 Hz. Electronic converters are required to couple flywheels to the electric grid.
The flywheel market is not mature and lacks of standardization. Brushless machines are preferred in flywheel applications, but there is still a great variety of machine topologies and system parameters discussed. The machine configurations may be classified into: induction, reluctance and permanent magnet machines.

3.1 Permanent Magnet machines

High coercitive materials have been developed and applied only for the last 20 years and the technology is still evolving. The rotor is shelf excited with Permanent Magnet (PM) excitation and allows high power density and efficiency as it lacks excitation losses (Gieras & Wing, 2002). These properties make PM machines preferred in many vehicular applications.

3.1.1 PM machine topologies

There is a great variety of permanent magnet arrangements to increase the magnetic field in the airgap, to obtain a sinusoidal distribution and to reduce eddy current losses in the magnets that may lead to reduction in performance and permanent demagnetization.
Regarding the flux path, most common types of machines have radial or axial flux configurations. Other topologies have been described without much widespread as conical, transversal or spherical. Magnets may be surface mounted or internal mounted on the rotor surface. The magnets are mounted on the rotor in different ways. Axial-flux machines usually have their magnets mounted on the surface of the rotor, while radial-flux machines may have the magnets either surface mounted or internal mounted (Kolehmainen & Ikäheimo, 2008). Internal mounted magnet machine properties vary with the geometry and configuration of the rotor. A magnetic material conducts the magnetic flux so the magnets are isolated from the harmonics produced by the stator. The iron bridges may be mechanized to obtain a sinusoidal magnetic flux distribution and produces a significant saliency. The saliency affects the performance of electric motors as lead to higher synchronous reactance in the direct axis (X_{sd}) than in the quadrature axis (X_{sq}). Iron bridges between and over the magnets produce a leakage in the magnetic flux, despite of the complexity of the arrangement.
The differences in geometry between surface mounted and Internal mounted magnets are clearly shown in Fig. 5.

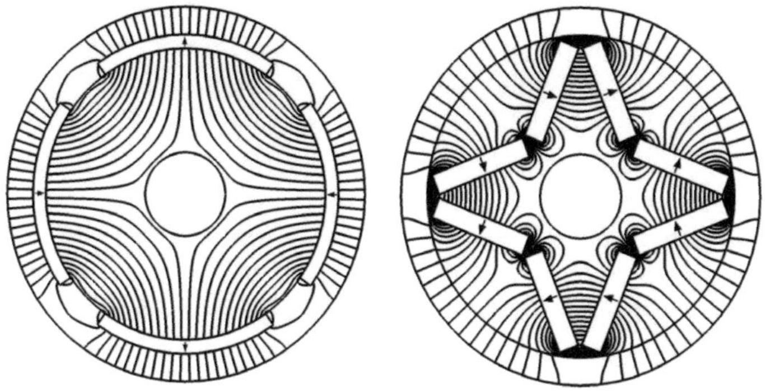

Fig. 5. Rotors with surface mounted (left) and internal mounted (right) magnets with the magnetic field lines induced in the airgap and in the rotor steel.

3.1.2 Halbach PM array

A special magnet configuration, both for axial and radial flux machines is the Halbach arrangement. For the ideal Halbach array, the magnets are combined in such a way that the magnetic field intensity is cancelled on one side of the array. With the Halbach magnet array no magnetic back-iron is needed and higher specific torques may be achieved. The simplest Halbach array configuration, presented in Fig. 6, conbines radial and azimutal magnets.

Fig. 6. Halbach array configuration. Magnetic potential in a radial magnet array (top), azimutal magnet array (middle) and composition of both in a Halbach array (bottom).

Electric machines with Halbach PM arrays have comparable performance as machines with a magnetic back yoke (Ofori-Tenkorrang & Lang, 1995) and an intrinsic sinusoidal magnetic field distribution in the airgap. The mass inertia in machines without iron back yoke is also lower, but the dinamic performnance is nor relevant for machines coupled to a high moment of inertia flywheel.

The size of high speed machines is usually limitd by the mechanical strenght of the magetic iron in the back yoke. Halbach arrays allow machine topologies without back yoke and self magnetic shielded, and some authors claim to be the best solution high speed machies for flywheel aplications (Post et al., 1993).

3.1.3 Coreless machines

Losses in the iron core of electric machines increase dramatically with electric frequency. This is the reason why coreless machine topologies are raising interest for high speed machines. With the development of new permanent magnet machines high magnetic fluxes may be achieved in the airgap of electric machines, and expected to reach higher values than traditional slotted machines (Santiago & Bernhoff, 2010).

Traditionally, stator windings are placed in laminated steel slots. The stator teeth reduce the airgap and therefore the magnetic reluctance in the magnetic circuit. Lower magnetic reluctance leads to less magnetic material, more compact designs and higher power density. There is a limit in the reduction of the airgap. Without considering the technical feasibility of construction, smaller airgaps have also some disadvantages. The magnetic flux distribution in the airgap becomes squared and cogging torque increases. Losses in the stator teeth also increase with smaller airgaps due to the increase in the harmonic content in the magnetic flux density.

There are two stator configurations without teeth. In slotless machines the windings are directly placed over the stator yoke. The magnetic flux path that goes through the stator back yoke has a substantially less density than in stator teeth (Wallmark et al., 2009).

In the ironless or coreless configuration, the back iron yoke rotates simultaneously with the rotor, so the magnetic circuit does not produce hysteresis or eddy current losses. The coreless stator reduces the iron loss, especially at high-speed operation (Ooshima et al., 2006). An ironless axial flux and a radial flux machine with an outer rotor configuration are presented in Fig. 7.

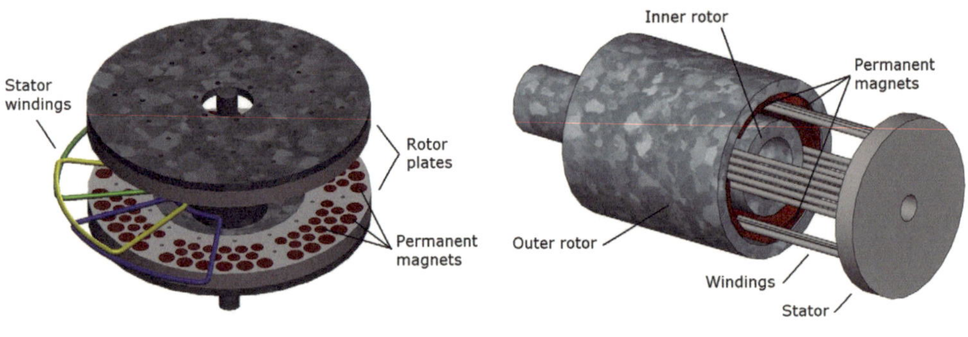

Fig. 7. Ironless axial and radial flux machines with an outer rotor configuration.

3.1.4 PM machine control

High coercitive materials such as Neodymium Iron Boron magnets have a very low magnetic permeability, close to air. This leads to very low inductance in the windings, especially for slotless machines. Low inductance machines require current control to reduce current ripple (Su & Adams, 2001).

For high performance motion control applications, the closed loop control with vector control should be incorporated to achieve high dynamic performance in position, speed and torque control (Jahns, 1997). However, when high dynamic performance is not a demand, simple V/f control strategies may be sufficient to obtain the required control performance.

3.1.4.1 Mathematical Model of a PM motor drive

The DQ transformation expresses the three-phase stationary coordinate system into the *d-q* rotating coordinate system (Low et al., 1995). Permanent magnet synchronous motors (PMSM) are described by a multivariable, coupled and nonlinear equations. The *d-q* transformation is used to transform these nonlinear equations into a simplified linear state model. The voltage equations of the PMSM in the rotating reference frame are:

$$v_d = R \cdot i_d + L_d \cdot \frac{di_d}{dt} - \omega \cdot L_q \cdot i_q \quad (2)$$

$$v_q = R \cdot i_q + L_q \cdot \frac{di_q}{dt} + \omega \cdot L_d \cdot i_d + \omega \cdot \lambda_r \quad (3)$$

The electromagnetic torque can be written as

$$T_e = \frac{3}{2} \cdot \frac{p}{2} \cdot \left[\lambda \cdot i_q + \left(L_d - L_q \right) \cdot i_q \cdot i_d \right] \quad (4)$$

where v_d, v_q, i_d, and i_q are the stator voltages and currents, respectively. R is the stator resistance, L_d and L_q are the d-q axis stator inductances, respectively; λ_r is the rotor flux, T_e is the electromagnetic torque and p is the number of poles.
The electromechanical equation of a PMSM is given by:

$$\frac{p}{2} \cdot \left(T_e - T_l \right) = J \cdot \frac{d}{dt} \cdot \omega + B \cdot \omega \quad (5)$$

where T_l, ω, J and B represent the load torque, the electrical rotor speed, the inertia and the friction coefficient of the motor, respectively.

3.1.4.2 Scalar V/f Control

The simplest way to control a PMSM for variable speed applications is through the open loop scalar control. It is used in applications where information about the angular speed is not needed. It is suitable for a wide range of drives as it ensures robustness at the cost of reduced dynamic performance. The supply voltage frequency is changed independently from the shaft response (position and angular speed). The magnitude of the supply voltage is changed according to the frequency in a constant ratio. Then the motor is in the condition where the magnetic flux represents the nominal value and the motor is neither overexcited nor underexcited. The main advantage of this simple method is the absence of a position sensor. The control algorithm does not need information about the angular speed or actual rotor position. On the contrary, the big disadvantages are the speed dependence on the external load torque, mainly for Induction Machines, and limited dynamic performances (Perera et al., 2002). Despite of its simplicity, scalar V/f control is used in flywheel applications (Sun et al., 2009).

The machine rotational speed ω varies proportionally with the frequency of the input signal, f_s, as follows:

$$\omega = \frac{2\pi \cdot f_s}{pp} \tag{6}$$

where *pp* is the number of pole pairs of the machine.
The magnetic flux can, if the stator resistance is neglected, be expressed as:

$$\lambda_s = \frac{V_\varphi}{\sqrt{2}\pi \cdot f_s \cdot N \cdot k_w} = Const.\frac{V_\varphi}{f_s} \tag{7}$$

So, in order to avoid variations in the stator flux (which could cause the motor to be overexcited or under excited), the voltage-to-frequency ratio is kept constant, hence the name V/f control.
If the ratio is different from the nominal one, the motor will become overexcited. This means that the magnetizing flux is higher than the constant ratio V/f, or underexcited, which happens because voltage is kept constant and the value of the stator frequency is higher than the nominal one.

3.1.4.3 Vector Control

Vector control (Field Oriented Control) of AC machines, as a novel approach in electrical drives, provides very good performance in dynamic responses in comparison with the scalar control. Vector control eliminates almost all the disadvantages of constant V/f control. The main idea of this method is based on controlling the magnitudes and angles of the space vectors.
Vector control of PMSM allows, by using *d-q* components, separating closed loop of both flux and torque (Stulrajter et al., 2007), hence, achieving a similar control structure to that of a separately excited DC machine. The electromagnetic torque can be expressed in *d-q* components according to nonlinear model of PMSM, as seen in equation 4. The torque depends on the rotor type and its inductances L_d, L_q and on permanent magnets mounted on the rotor.
The surface mounted (non-salient) PMSM, it can be taken that $L_d = L_q$ and the maximum torque per ampere for this machine is obtained by making $i_d = 0$, or, in other words, by maintaining the torque angle at 90° what produces a maximum quadrant current i_q.
It follows from equation 4 that if a non-salient machine is considered the electromagnetic torque can be expressed as

$$T_e = \frac{3}{2} \cdot \frac{p}{2} \cdot \lambda \cdot i_q \tag{8}$$

Vector control structures for a wide variety of PMSM drivers have the same characteristic. The most popular control technique is the cascaded one using classical techniques to achieve torque, speed and position control in PMSM motion control system, as seen in Fig. 8.
Fig. 8 shows a closed speed feedback loop around the inner torque/current loop. The torque request is generated by the speed controller and, by keeping i_d equal to zero, the phase stator current will be placed in the quadrature axis and the maximal driving torque will be achieved.

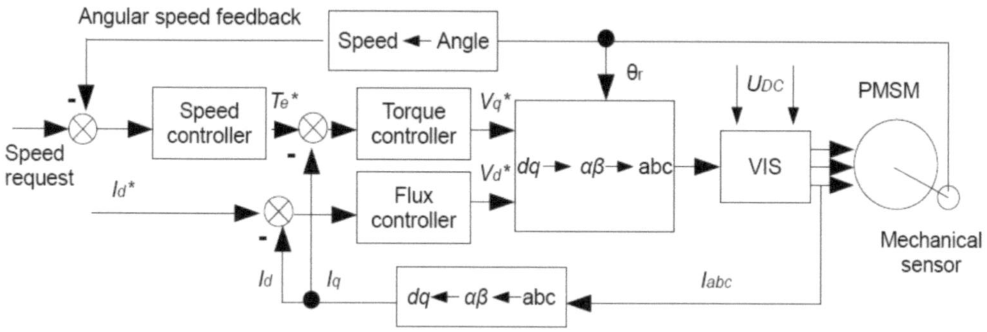

Fig. 8. Typical cascaded control structure for PMSM drivers.

3.1.4.4 Variable geometry for variable speed operation

Magnetization in permanent magnets is constant and can not be used as a control parameter as synchronous machines regulate the field current. Therefore the voltage increases linearly with speed, in absence of magnetic saturation. Variable geometry topologies have been proposed to operate permanent magnets in a wide speed range with constant back e.m.f.

There are two different variable geometries strategies reported for variable speed operation. The solution proposed in (Javadi & Mirsalim, 2010) is based on a double-stator structure with variable stator geometry. The concept has been applied in an axial flux generator with a three stator and a coreless double-stator structure; one is stationary and the other rotate to achieve field weakening.

The field weakening may be achieved by increasing the airgap. This field weakening strategy may only be applied in axial flux machines, as the radius of radial flux machines is inherently constant. In axial flux machine with one stator and one rotor configuration, a mechanism that separates the rotor from the stator as the speed increases may be implemented. With the same excitation in the rotor, a higher airgap reduces the magnetic flux through the stator. The back e.m.f. amplitude is kept constant for a high range of speeds, as the electrical frequency increases linearly with the speed. An advantage in this system is that the efficiency is very high for a wide range of operational speeds. The hysteretic and eddy current losses are proportional to the square of the speed, but also to the magnetic flux. At high speed the frequency increases, but the magnetic field is reduced, counteracting this effect. Efficiencies of 98% at 10.000 rpm with this system have been reported (Nagaya et al., 2003).

3.2 Induction machines

About 65% of the worlds electricity production in the world is consumed in induction motors. They are ussually preferred because they are inexpensive, require little maintenance and are reliable. Asynchronous machines have a very mature and standarized technology; EPA in the US and CEMEP in the European Union have a general efficiency clasification system. The result of adapting these directives will lead to an increase in the efficiency and a shift to high efficiency machines with a signifficant impoct in the market for the next years (Chitroju, 2009). The equivalent circuit of an induction machine is presented in Fig. 9 where the influence of the slip is clearly seen.

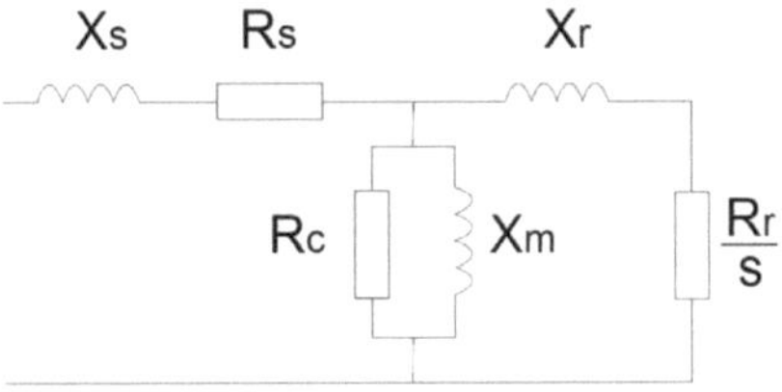

Fig. 9. Excitation system description.

Variable frequency AC drives are used for variable speed operation. The inverter allows a great flexibility of power and speed range but the power electronic rating requiered increases the equipment costs for high power aplications. Due to the system simplicity, asynchronous machines are used in stationary flywheel energy storage low power applications (Cheng et al., 2008), (Kato et al., 2009).
A strategy to reduce the power rating of the power electronics consist of using two sets of windings in the stator. One set would be directly connected to the grid and the other would be driven by an inverter. This is the idea behind Brushless Doubly Fed Induction Machine (BDFIM) topology proposed for a regular 2:1 speed range ratio in a flywheel operation (Tazil et al., 2010).

3.2.1 Induction machines control

One way of controlling AC motors for variable speed applications is through the open loop scalar control, which represents the most popular control strategy of squirrel cage AC motors.
Most of the concepts in control estimation for permanent magnets synchronous motors are also applicable to induction motor drives. Open loop scalar control is broadly used in induction motors drives (Finch, 1998), (Luo et al., 2007), (Srilad et al., 2007), however, its importance is diminishing because of the superior performance of vector controlled (or field oriented controlled) drives (Khambadkone & Holtz, 1991), (Kim et al., 1986), (Rowan et al., 1982), (Xu et al., 1988).
Scalar- and vector-controlled drives have already been discussed in Section 3.1.4. An advanced scalar control technique, know as direct torque and flux control (DTC) (Habetter, 1992) was introduced in the mid-1980s, being claimed to have nearly comparable performance with vector-controlled drives. DTC has recently been introduced in commercial induction motor drives thus creating a wide interest.

3.2.1.1. Direct Torque and Flux Control

Direct Torque Control (DTC) uses an induction motor model to predict the voltage required to achieve a desired output torque (Takahashi & Noguch, 1986). Differently from vector control, stator flux and output torque are estimated by using only current and voltage measurements according to equations 9 and 10:

$$\overline{\lambda}_S = \left(\overline{V}_S - r_S \cdot \overline{I}_S\right) \cdot dt \tag{9}$$

$$T_{em} = \frac{3}{2} \cdot \frac{p}{2} \cdot \left(\overline{\lambda}_S \times \overline{I}_S\right) \tag{10}$$

where $\bar{\lambda}_S$ is the stator flux vector, T_{em} is the produced torque, p is the number of poles. $\bar{V}_S$, $\bar{I}_S$ and r_s are the stator voltage, current and resistance, respectively.
Combining equations 9 and 10 to the equations that describe the equivalent circuit of an induction motor, expressions for the change in torque and flux are obtained. These equations can be solved to find the smallest voltage vector, $\bar{V}_S$, required to drive both the torque and flux to the demand values.
Fig. 10 shows the schematic of the basic functional blocks used to implement Direct Torque Control:

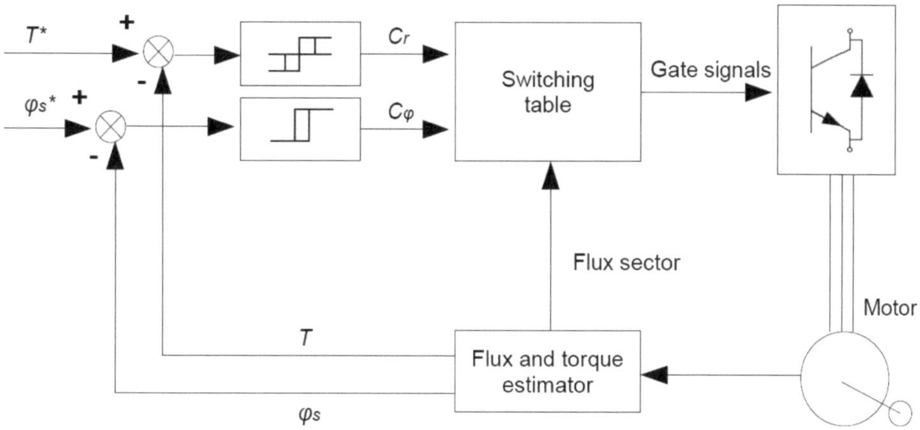

Fig. 10. Basic Direct Torque Control scheme.

Latest research on DTC has focused on decreasing the torque ripple and obtaining faster transient response to the step changes in torque during start-up (Cadasei & Serra, 2002), (El afia et al., 2005). Also, the combination of DTC and intelligent techniques such as fuzzy logic or artificial neural network has been attracting the attention of many scientists from all over the world (Toufouti et al., 2006), (Toufouti et al., 2007).

3.2.1.2 Brushless Doubly Feed Induction Machines (BDFIM)

BDFIM machines are gaining attention in wind power generation as limmited variable speed is required. The BDFIM has similar rotor as the singel feed traditional induction machine. It is a solid piece of laminated steel with conducting bars, but instead been shortcircuited forming a cage, the end winding conections form poles as presented in Fig. 11. The stator structure do not differe from the induction machine. The difference is that the BDFIM has two sets of insulated stator windings of different pole numbers. One primary winding (or power winding) is grid connected and the secondary winding (or control winding) is driven by a converter that regulated the frequency. The machine speed is the composition of the primary and secundary winding frequencies.
The power in the BDFIM is partially driven by the secundary winding, but most of it flows directly from the rotor to the power winding, reducing the power electronic rating till only 25% of the requierements of a single feed induction machine (Klempner & Kerszenbaum, 2004). The BDFIM has been proposed and implemented for flywheel applications with promising results (Wu et al., 2009).

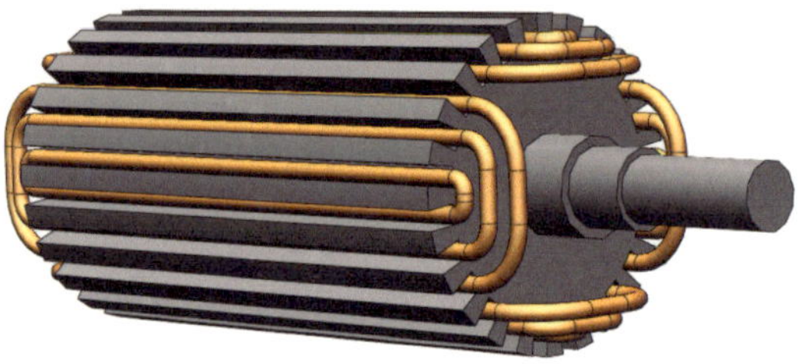

Fig. 11. Four pole rotor configuration of a BDFIM.

3.3 Reluctance machines

The main characteristic of reluctance machines is that the rotor is built with salient poles. The rotor lacks excitation and the torque is produced solely by the difference between the direct axis and quadrature axis synchronous reactance. The power is therefore obtained by the second term of equation 1. A figure of merit for the synchronous reluctance machine is the ratio of *d* to *q* axis inductance.

Reluctance machines have some favourable characteristics for flywheel energy storage systems. It lacks excitation currents that allow low idle losses. The rotor is robust and allows high speeds operations. It has great acceptance in UPS systems and other stationary applications as stand alone flywheels to handle voltage sags and power disruptions that last less than 5 seconds (Park et al., 2008). This reduces the number of charge/discharge cycles and increases the lifespan of the battery pack. The equivalent circuit of a reluctance machine is presented in Fig. 12.

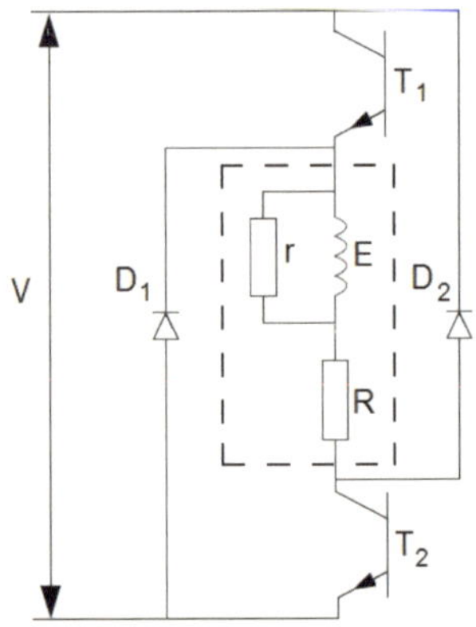

Fig. 12. One phase equivalent circuit with switching elements of a reluctance motor.

Reluctance machines may be divided into switched-reluctance machines (SRM) and synchronous reluctance machines (Syncrel).

The Syncrel has distributed windings in the stator, similar to synchronous and inductance machines. It lacks of any source of flux on the rotor, and therefore the power density is lower than in synchronous or induction machines. To increase the power density and efficiency, permanent magnet may be placed in the rotor. This hybrid type of machine is called permanent magnet assisted synchronous reluctance motor.

The SRM has concentrated windings and a saliency structure both in rotor and stator, that is why there are also called doubly salient variable-reluctance machines. The structure of the machines is similar to the stepper motor. There are many combinations in the number of stator phases and rotor poles. The simplest consists of only one phase with a considerable cogging torque. The most common configuration has a four-phase and eight rotor poles and six stator poles configuration. At least three phases are required for a four quadrants operation (both motoring and generating) (Rashid, 2007).

Reluctance machines may have two sets of independent windings in the stator. The Double Feed Reluctance Machine (DFRM) has not been commercially developed yet, but is has the same potentially advantages as the DFIM. The DFRM has two sinusoidal distributed stator windings as the DFIM discussed in 3.2.2. The power electronics require lower power ratings than single feed machines and also allow variable speed operation while the power winding is directly coupled to the grid frequency (Valenciaga & Puleston, 2007). The main drawback of this technology is the low torque per volume, lower than an equivalent synchronous reluctance (Syncrel) or a cage induction machine (IM) (Jovanovic, 2009).

4. Conclusion

Energy storage development is essential if intermittent renewable energy generation is to increase. Pumped hydro, CAES and flywheels are environmentally friendly and economical storage alternatives that required electric motor/generators. The popularization of power electronics is relatively new and therefore the technology is still under development. There is not a clear winner when comparing technologies and therefore the optimal alternative depends on the specific requirements of the application.

In this chapter the main electric machine topologies for energy storage are presented. The discussion is focused on the applicability and also on the latest research threads and state of the art.

5. References

Acarnley, P.P.; Mecrow, B.C.; Burdess, J.S.; Fawcett, J.N.; Kelly, J.G. & Dickinson, P.G. (1996), Design principles for a flywheel energy store for road vehicles, *IEEE Trans. on Industry Applications,* Vol. 32 , Issue: 6, pp. 1402 - 1408, 1996.

Boger, M. S.; Wallace, A. K. & Spee, R. (1996), Investigation of Appropriate Pole Number Combinations for Brushless Doubly Fed Machines Applied to Pump Drives, *IEEE Trans. On Industry Applications,* Vol. 32, Issue 1, pp. 189-194, 1996.

Carrillo, C., Feijóo, A. & Cidrás J. (2009), Comparative study of flywheel systems in an isolated wind plant, *Renewable Energy,* Vol. 34, Issue 3, pp. 890-898,March 2009.

Casadei, D. & Serra, G. (2002), Implementation of direct Torque control Algorithm for Induction Motors Based on Discrete Space Vector Modulation, *IEEE Trans. Power Electronics,* Vol. 15, No. 4, 2002.

Changda, S.; Guobiao G. & Shunchao, Y. (1998), Evaporative cooling technology more suitable for the Three Gorges' generators, *Proceedings of POWERCON '98. International Conference on Power System Technology,* pp 1025-9, Vol. 2, 1998.

Cheng, M., Kato, S., Sumitani, H., & Shimada, R. (2008), A novel method for improving the overload capability of stand-alone power generating systems based on a flywheel induction motor, *Proceedings of the IEEE Annual Power Electronics Specialists Conference*, pp. 3677-3683, 2008.

Chitroju, R. K. S. (2009), Improved Performance Characteristics of Induction Machines with Non-skewed Asymmetrical Rotor Slots, *Licentiate thesis, Royal Institute of Technology (KTH)*, Stockholm, Sweden, 2009. ISBN 978-91-7415-453-5.

El afia, A., Khafallah, M., Cheriti, A., Elmoussaoui, B. & Saad, A. (2005), A simple direct torque fuzzy control of induction motor using space vector modulation, *EPE Journal*, Vol. 15, No. 2, pp. 25-30, 2005.

Finch, J. W. (1998), Scalar and vector: a simplified treatment of induction motor control performance, *Proceedings of the IEE Colloquium on Vector Control Revisited, IEE Digest*, Vol. 1998, Issue 199, pp. 2/1-4, 1998.

Fuchs, E. F. & Rosenberg L. T. (1974), Analysis of an alternator with two displaced stator windings, *IEEE Trans. Power Application Systems*, Vol. PAS–93, pp. 1776–1786, 1974.

Gieras, J. F. & Wing, M. (2002), *Permanent Magnet Motor Technology Revised*, CRC, 2nd edition, January 15, 2002.

Habetler, T. G., Profumo, F., Pastorelli, M. & Tolbert, L.M. (1992), Direct Torque Control of Induction Machines using Space Vector Modulation, *IEEE Trans. Industry Applications*, Vol. 28, No. 5, 1992.

IEEE Std 421.1-2007, IEEE Standard Definitions for Excitation Systems for Synchronous Machines.

Jahns, T. M. (1997), Variable Frequency Permanent Magnet AC Machine Drives, Chapter 6 in *Power Electronics and Variable Frequency Drives, Technology and Applications*, Ed. IEEE Press, ISBN o-7803-1061-6, 1997.

Javadi, S. & Mirsalim, M. (2010), Design and Analysis of 42-V Coreless Axial-Flux Permanent-Magnet Generators for Automotive Applications, *IEEE Trans. on Magnetics*, Vol. 46, Issue 4, pp. 1015-1023, April 2010.

Jovanovic, M. (2009), Sensored and sensorless speed control methods for brushless doubly fed reluctance motors, *IET Electric Power Applications*, Vol. 3, Issue 6, pp. 503–513, November 2009.

Kato, S., Takaku, T., Sumitani, H. & Shimada, R. (2009), Development of voltage sag compensator and UPS using a flywheel induction motor and an engine generator, *Electrical Engineering in Japan*, Vol. 167, No. 1, pp. 74-81, 15 April 2009.

Khambadkone, A. M. & Holtz, J. (1991), Vector controlled induction motor drive with a self-comissioning scheme, *IEEE Trans. on Industrial Electronics*, Vol. 38, p 322-327, 1991.

Kim, C. (2002), Performance Analysis for the Modified Excitation System of Synchronous Machine Connected to HVDC System, *KIEE International Transaction on Electrical Machinery and Energy Conversion Systems*, Vol. 2B, No. 3, 2002, pp. 109-114, 2002, ISSN 1598-2602.

Kim, Y. R., Sul, S. K. & Park, M. H. (1986), Speed sensorless vector control of induction using extended Kalman filter, *IEEE Trans. on Industrial Applications*, Vol. 30, pp 1225-1233, 1986.

Klempner, G. & Kerszenbaum, I. (2004). Operation and maintenance of large turbo-generators, *IEEE Press Series on Power Engineering*, ISBN 0-471-61447-5, New Jersey.

Kolehmainen, J. & Ikäheimo, J. (2008), Motors With Buried Magnets for Medium-Speed Applications, *IEEE Trans on Energy Conversion*, Vol. 23, No. 1, March 2008.

Laughton, M.A. & Warne, D.F, (2003), *Electrical Engineer's Reference Book*, 16th Edition, Chapter 28, ISBN 13: 9780750646376, Elsevier, 2003.

Low, K. S., Rahman, M. F. & Lim, K. W. (1995), The d-q transformation and feedback linearization of a permanent magnet synchronous motor, *Proceedings of International Conference on Power Electronics and Drive Systems*, Vol. 1, pp. 292-296, 1995.

Lundin U & Wolfbrandt A (2009), Method for Modeling Time-Dependent Nonuniform Rotor/Stator Configurations in Electrical Machines, *IEEE Trans. on Magnetics*, Vol. 45, Issue 7, pp. 2976-2980, 2009.

Luo, H., Wang, Q., Deng, X. & Wan. S. (2007), A novel V/f scalar controlled induction motor drives with compensation based on decoupled stator current, *IEEE International Conference on Information Technology*, pp. 1989-94, 2007.

Mamikoniants, L.G., Shakaran, Y.G.; Vinitsky, Y.D., Kauffmann, J.M. & Barbier, G. (1999). Asynchronized synchronous machine application in the power systems - operational experience and prospects, *Proceedings of International Electric Machines and Drives Conference IEEE IEMDC'99*, pp. 746 - 748, Seattle, Washington, USA, May 1999.

Nagaya, S., Kashima, N., Kawashima, H., Kakiuchi, Y., Hoshino, A. & Isobe, S. (2003), Development of the axial gap type motor/generator for the flywheel with superconducting magnetic bearings, *Physica C- Superconductivity and its Applications*, Vol. 392, pp. 764–768, 2003.

Ofori-Tenkorrang, J. & Lang, J.H. (1995), A Comparative Analysis of Torque Production in Halbach and Conventional Surface-Mounted Permanent-Magnet Synchronous Motors, *Proceedings of 13th IEEE Industry Applications Conference*, IAS 1995.

Ooshima, M., Kitazawa, S., Chiba, A., Fukao, T. & Dorrell, D.G. (2006), Design and analyses of a coreless-stator-type bearingless motor/generator for clean energy generation and storage systems, *IEEE Transactions on Magnetics*, Vol. 42, No. 10, pp. 3461-3, Oct. 2006.

Park, J., Kalev, C. & Hofmann, H.F. (2008), Control of High-Speed Solid-Rotor Synchronous Reluctance Motor/Generator for Flywheel-Based Uninterruptible Power Supplies. *IEEE Trans. on Industrial Electronics*, Vol. 55, No. 8, pp. 3038-46, Aug. 2008.

Perera, P. D. C., Blaabjerg, F., Pedersen, J. K. & Thogersen, P. (2002), A sensorless, stable V/f control method for permanent-magnet synchronous motor drives, *Proceedings of Applied Power Electronics Conference and Exposition*, Vol. 1, pp. 83 – 8, 2002.

Post, R.F.; Fowler, T.K. & Post, S.F., (1993), A High-Efficiency Electromechanical Battery, *Proceedings of the IEEE*, Vol. 81, No. 3, pp. 462-74, March 1993.

Rashid, M. H., (2007), Motor drives, Chapter 33 in: *Power Electronics Handbook: Devices, Circuits, and Applications*, Second Edition, ISBN-13: 9780120884797, Academic Press, 2007.

Rowan, T. M., Kerkman, R. J. & Leggate, D. (1982), A simple on-line adaptation for indirect field orientation of an induction machine, *Proceedings of IEEE IAS Annual Meeting Conference*, pp. 604-609, 1982.

Santiago, J. & Bernhoff, H. (2010), Comparison between axial and radial flux PM coreless machines for flywheel energy storage, *Journal of Electrical Systems*, Issue 2, Vol. 6, (June 2010), ISSN 1112-5209.

Srilad, S., Tunyasrirut, S. & Suksri, T. (2007), Implementation of a scalar controlled induction motor drives, *Proceedings of 2006 SICE-ICASE International Joint Conference (IEEE Cat. No. 06TH8879)*, p 6, 2007.

Stulrajter, M., Hrabovcova, V. & Franko, M.(2007), Permanent magnet synchronous motor control theory, *Journal of Electrical Engineering*, Vol. 58, No. 2, pp. 79-84, 2007.

Su, G. J. & Adams, D. J (2001), Multilevel DC link inverter for brushless permanent magnet motors with very low inductance, *Proceedings of IEEE Industry Applications Society Annual Meeting*, pp. 829-834, 2001.

Sun, X. D., Koh, K. H., Yu, B. G. & and Matsui, M., (2009), Fuzzy-Logic-Based V/f Control of an Induction Motor for a DC Grid Power-Leveling System Using Flywheel Energy Storage Equipment, *IEEE Trans. on Industrial Electronics*, Vol. 56, No. 8, 2009.

Takahashi, I. & Noguch, T. (1986), A new quick response and high efficient control strategy for an induction motor, *IEEE Trans. Industrial Applications*, Vol. 22, pp. 820-827, 1986.

Tazil, M., Kumar, V., Bansal, R.C., Kong, S., Dong, Z.Y. & Freitas, W. (2010), Three-phase doubly fed induction generators: an overview, *IET Electric Power Applications*, Vol. 4, No. 2, pp. 75-89, Feb. 2010.

Toufouti, R., Meziane, S. & Benalla, H. (2006), Direct Torque Control for Induction Motor Using Fuzzy Logic, *ACSE Journal*, Vol. 6, No. 2, June 2006.

Toufouti, R., Meziane, S. & Benalla, H. (2007), Direct torque control for induction motor using intelligent techniques, *Journal of Theoretical and Applied Information Technology*, pp. 35-44, 2007.

Touma-Holmberg, M. & Srivastava, K. (2004), Double winding, high-Voltage cable wound Generator: steady-State and fault analysis, *IEEE Trans. on Energy Conversion*, Vol. 19, No. 2, pp. 245-50, June 2004.

Valenciaga, F & Puleston, P. F. (2007), Variable Structure Control of a Wind Energy Conversion System Based on a Brushless Doubly Fed Reluctance Generator, *IEEE Trans. on Energy Conversion*, Vol. 22, No. 2, 2007.

Wallmark, O., Kjellqvist, P. & Meier, F. (2009), Analysis of axial leakage in high-speed slotless PM motors for industrial hand tools, *IEEE Trans. on Industry Applications*, Vol. 45, No. 5, pp. 1815-20, Sept.-Oct. 2009.

Wu, J., Wen, J., & Sun, H. (2009), A New Energy Storage System Based On Flywheel, *Proceedings of IEEE Power & Energy Society General Meeting (PES)*, pp. 1-6 , 2009, ISSN: 1944-9925.

Xu, X., De Doncker, R. & Novotny, D. W. (1988), A stator flux oriented induction machine drive, *Proceedings of IEEE Power Electronics Specialists Conference*, pp. 870-876, 1988.

2

Control of a DSTATCOM Coupled with a Flywheel Energy Storage System to Improve the Power Quality of a Wind Power System

Gastón Orlando Suvire and Pedro Enrique Mercado
Instituto de Energía Eléctrica – Universidad Nacional de San Juan
Argentina

1. Introduction

Wind power generation is considered the most economic viable alternative within the portfolio of renewable energy resources. Among its main advantages are the large number of potential sites for plant installation and a rapidly evolving technology. However, the lack of controllability over the wind and the type of generation system used cause problems to the electric systems. Among such problems are those produced by wind power short-term fluctuations, e.g., in the power quality and in the dynamics of the system (Slootweg & Kling, 2003; Ackermann, 2005; Suvire & Mercado, 2008; Chen & Spooner, 2001; Mohod & Aware; 2008; Smith et al., 2007). In addition, the reduced cost of power electronic devices as well as the breakthrough of new technologies in the field of electric energy storage makes it possible to incorporate this storage with electronic control into power systems (Brad & McDowall, 2005; Carrasco, 2006; Barton & Infield, 2004; Hebner et al., 2002). These devices allow a dynamic control to be made of both voltage and flows of active and reactive power. Therefore, they offer a great potential in their use to mitigate problems introduced by wind generation.

Based on the results obtained by analyzing different selection criteria, a Distribution Static Synchronous Compensator (DSTATCOM) coupled with a Flywheel Energy Storage System (FESS) has been proposed as the most appropriate system for contributing to the smoothing of wind power short-term fluctuations (Suvire & Mercado, 2007). A DSTATCOM is a fast-response, solid-state power controller that provides flexible voltage control at the point of connection to the utility distribution feeder for power quality improvements (Song & Johns, 1999). This device can exchange both active and reactive power if an energy storage system is included into the DC bus. FESSs store kinetic energy in a rotating mass, and they have been used as short-term energy storage devices. FESSs can be classified as low-speed flywheel (LS-FESS) and high-speed flywheel (HS-FESS). HS-FESSs are a newer technology and they provide better speeds of response, cycling characteristics and electric efficiencies than LS-FESS (Hebner et al., 2002; Andrade et al., 2007). All these characteristics enable the HS-FESS (FESS from now on), working with a DSTATCOM device, to mitigate voltage fluctuations and to correct power fluctuations of a wind power system. With these aspects in mind, it turns necessary to ponder the information stemming from models that simulate the dynamic interaction between the DSTATCOM/FESS device and power systems with wind generation. Such models allow performing the necessary preliminary studies before connecting the DSTATCOM/FESS to the grid. Many solutions are proposed and studied in the literature to compensate wind power fluctuations using a flywheel energy storage device (Boutot et al., 2002; Takahashi et al., 2005; Cimuca et al., 2004; Cárdenas et al., 2004). These solutions have been proposed mainly using LS-FESS and with simplified models of the device. The complete control to interact with wind power generation is not explained in detail in the analyzed literature.

The aim of this paper is to present a detailed model and a multi-level control of a DSTATCOM controller coupled with FESS to improve the integration of wind generators (WGs) into a power system. A model of a DSTATCOM/FESS device is proposed with all its components represented in detail. Moreover, the complete control for this device is suggested. This control implements a new approach based on multi-level control technique. To mitigate wind power fluctuations, the control includes three modes of operation of the DSTATCOM/FESS device, namely, voltage control, power factor correction, and active power control. Validation of models and control schemes is carried out through simulations by using SimPowerSystems of SIMULINK/MATLAB™.

2. Modelling of the DSTATCOM/FESS

In order to study the dynamic performance of the DSTATCOM/FESS controller, a model of the combined system is proposed that is depicted in Fig. 1. This model consists mainly of the DSTATCOM controller, the Interface converter and the FESS device.

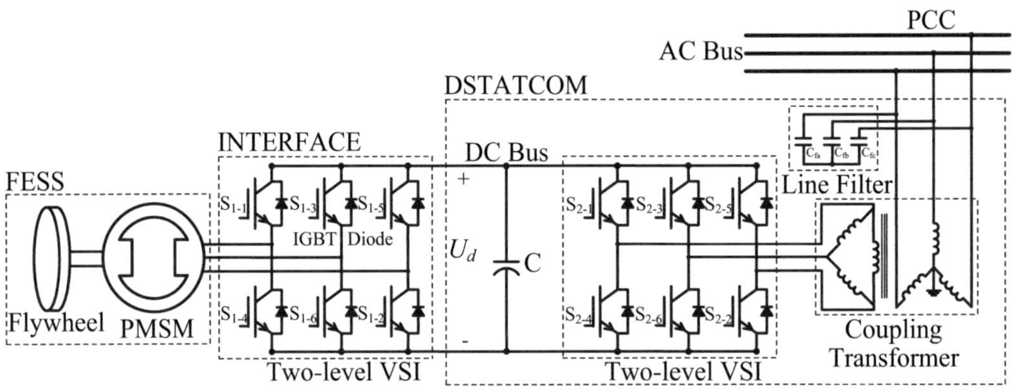

Fig. 1. Representation of the DSTATCOM/FESS controller

The DSTATCOM and the Interface use two-level VSIs. The commutation valves used are Insulated Gate Bipolar Transistors (IGBT) with anti-parallel diodes. The VSIs are modeled with detailed blocks of the switches and diodes, incorporated into the simulation program. The technique of sinusoidal pulse width modulation (SPWM) is used to obtain a sinusoidal voltage waveform. In order to reduce the disturbance produced on the distribution system by the high-frequency switching harmonics generated by the SPWM control, a low pass sine wave filter is used.

The energy stored by a FESS is calculated by using (1).

$$\Delta E = \frac{1}{2} J \left(\omega_{\max}^2 - \omega_{\min}^2 \right) \tag{1}$$

where ΔE is the energy stored by the flywheel, ω_{max} and ω_{min} are, respectively, the maximum and minimum operation speed of the flywheel, and J is the moment of inertia of the flywheel.

The exchange of power between the flywheel and the Interface is made by using a Permanent Magnet Synchronous Machine (PMSM). The PMSM is modeled with a detailed block included in the simulation program and with parameters obtained from the manufacturer data sheets (Beacon Power, 2009; Flywheel Energy Systems, 2009; Urenco Power Technologies, 2009). The flywheel is modeled as an additional mass coupled to the rotor shaft of the PMSM (Samineni et al., 2006).

3. DSTATCOM/FESS control

The control proposed for the DSTATCOM/FESS device is divided into two parts, the DSTATCOM control and the FESS control. For each part, a multi-level control scheme is suggested. This scheme has its own control objectives for each level. In this way, a system of complex control is divided into several control levels, which are simpler to design (Xie et al., 2002; Molina & Mercado, 2004). Both parts of the multi-level control scheme, i.e., the DSTATCOM and the FESS, are divided into three quite distinct levels: external, middle and internal level, shown in simplified way in Fig. 2.

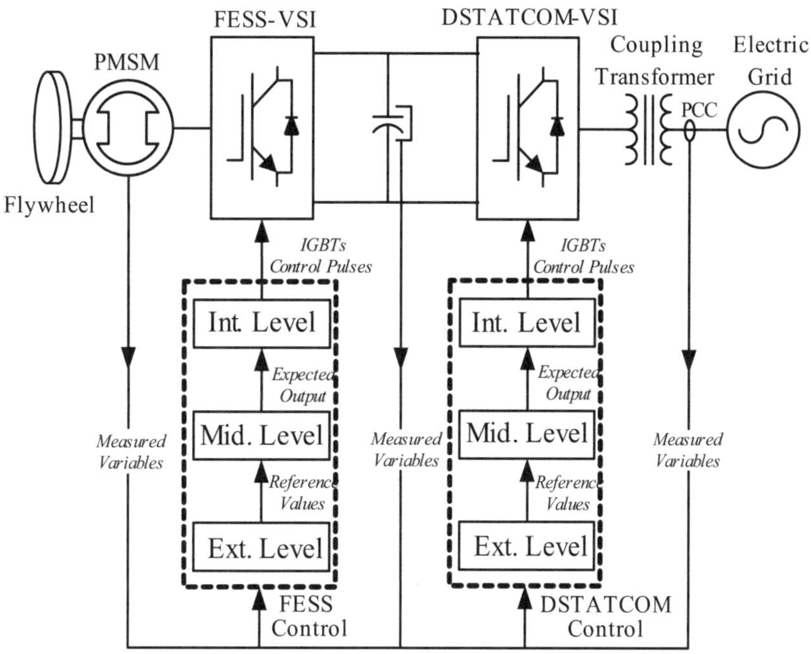

Fig. 2. Structure of the multi-level control of the DSTATCOM/FESS

3.1 DSTATCOM control

Each control level of the DSTATCOM has certain functions. The external level is responsible for determining the active and reactive power exchange between the DSTATCOM and the utility system. The middle level control allows the expected output to dynamically track the reference values set by the external level. The internal level is responsible for generating the switching signals for the valves of the VSI of the DSTATCOM. The control algorithm of the DSTATCOM with all its parts in detail is shown in Fig. 3.

Control is performed with the synchronous-rotating dq reference frame. The coordinate system is defined with the d-axis always coincident with the instantaneous voltage vector ($u_d = |u|$, $u_q = 0$). Consequently, the d-axis current component contributes to the instantaneous active power and the q-axis current component represents the instantaneous reactive power.

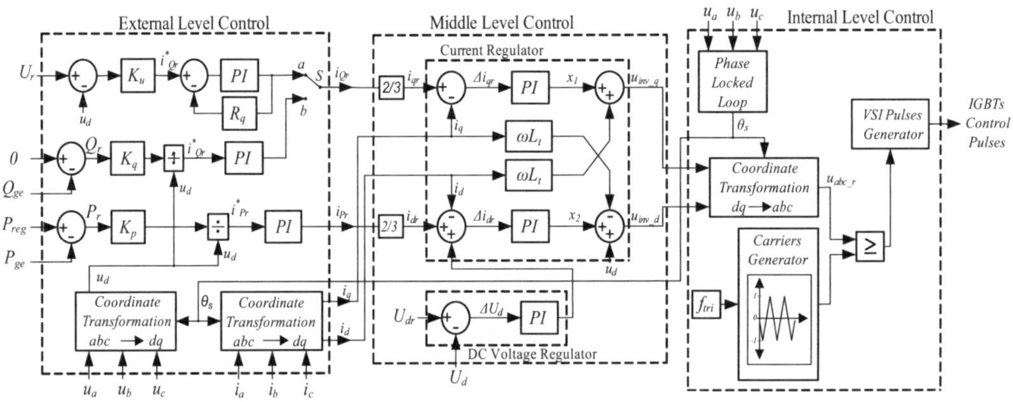

Fig. 3. Multi-level control scheme of the DSTATCOM device

External level control

The external level control scheme proposed (left side in Fig. 3) is designed for performing three major control objectives, namely, the voltage control mode (VCM), which is activated when switch *S* is in position *a*, the power factor control mode (PFCM), activated in position *b*, and the active power control mode (APCM), which is always activated.

The VCM consists in controlling the voltage at the PCC (Point of Common Coupling) of the DSTATCOM through the modulation of the reactive component of the output current. To this aim, the instantaneous voltage at the PCC (u_d) is computed by using a synchronous-rotating orthogonal reference frame and is then compared with a reference voltage (U_r). A voltage regulation droop (or slope) R_q is included in order to allow the terminal voltage of the DSTATCOM to vary in proportion with the compensating reactive current.

In the PFCM, the reactive power reference (Q_r) is set to the measured value of the reactive power of wind generation (Q_{ge}). In this way, all the reactive power required by the WG is provided and thus the WG-DSTATCOM/FESS system is able to maintain the unity power factor. A standard PI compensator is included to eliminate the steady-state error in the reactive current reference computation.

The APCM allows controlling the active power exchanged with the electric system. The computation of the reference active power (P_r) depends on the active power value injected by the wind generation. This value is the difference between the regulation power desirable (P_{reg}) and the active power measured from the WG (P_{ge}). The P_{reg} is the active power that needs to be delivered to the electric system by the WG-DSTATCOM/FESS system. A standard PI compensator is also included to eliminate the steady-state error in the active current reference computation.

Middle level control

This block has two main parts, the DC voltage regulator and the current regulator. A functional simplified scheme of this control level is shown in the central part of Fig. 3.

The dynamic equations governing the power instantaneous transfer between the DSTATCOM and the electrical network are given by (2).

$$\frac{d}{dt}\begin{bmatrix} i_q \\ i_d \end{bmatrix} = \begin{bmatrix} -\omega & -R_t / L_t \\ -R_t / L_t & \omega \end{bmatrix}\begin{bmatrix} i_d \\ i_q \end{bmatrix} + \frac{1}{L_t}\begin{bmatrix} u_{inv-q} \\ u_{inv-d} - |u| \end{bmatrix} \tag{2}$$

where R_t and L_t are, respectively, the resistance and equivalent leakage inductance of the coupling transformer of the DSTATCOM.

A control methodology to obtain a decoupled control of the current components, i_d and i_q, is derived from (2). To achieve this objective, two appropriate control signals x_1 and x_2 are introduced. If $i_qR_t/L_t = x_1$ and $i_dR_t/L_t = x_2$, and (2) is worked and these variables introduced; then (2) results in (3).

$$\frac{d}{dt}\begin{bmatrix} i_q \\ i_d \end{bmatrix} = \begin{bmatrix} 0 & -R_t / L_t \\ -R_t / L_t & 0 \end{bmatrix}\begin{bmatrix} i_d \\ i_q \end{bmatrix} + \begin{bmatrix} x_1 \\ x_2 \end{bmatrix} \tag{3}$$

As can be noticed from the equation above, i_d and i_q respectively respond to x_1 and x_2 with no cross-coupling. Conventional PI controllers with proper feedback from the DSTATCOM/FESS output current component are used to obtain the decoupling condition.

In addition, the AC and DC sides of the DSTATCOM are related by the power balance between the input and the output as described by (4).

$$P_{AC} = \frac{3}{2}\left(u_{inv\_d}i_d + u_{inv\_q}i_q\right) = -CU_d\frac{dU_d}{dt} - \frac{U_d^2}{R_{pd}} = P_{DC} \tag{4}$$

where R_{pd} is the loss resistance of the VSI and U_d is the DC voltage. Considering $u_{inv\_d} = k_{inv}\cos\alpha U_d$ and $u_{inv\_q} = k_{inv}\sin\alpha U_d$, with $k_{inv} = m_a a_t/2$, (m_a modulation index, $a_t = n_1/n_2$: voltage ratio of the coupling transformer) and α is the phase-shift between the converter output voltage and the grid AC voltage; (4) may be rewritten as:

$$\frac{dU_d}{dt} = -\frac{3}{2}\frac{1}{C}k_{inv}\cos\alpha\, i_d - \frac{3}{2}\frac{1}{C}k_{inv}\sin\alpha\, i_q - \frac{U_d}{CR_{pd}} \tag{5}$$

Another PI compensator which allows eliminating the steady-state voltage variations at the DC bus is used by forcing a small active power exchange with the electric grid.

Internal level control

A basic scheme of the internal level control of the DSTATCOM is shown on the right side of Fig. 3. This level is mainly composed of a line synchronization module and a three-phase PWM firing pulses generator for the DSTATCOM VSI. The line synchronization module consists mainly of a phase locked loop (PLL) (Bose, 2002). The three-phase firing pulses generator produces both a frequency triangular wave (f_{tri}) and the firing pulses for each IGBT of the VSI by comparing this triangular wave with the desired reference three-phase voltage, u_{abc_r}.

3.2 FESS control

The FESS control is carried out through the control of the Interface-VSI. By establishing a three-phase voltage of controllable amplitude and phase with the VSI, the PMSM can work as a motor storing energy or as generator delivering energy. In a way similar to the DSTATCOM control, each control level has to perform certain functions. The external level is responsible for determining the power exchange between the DC bus of the DSTATCOM and the FESS so as to fulfil the power requirements imposed by the DSTATCOM. The middle and internal levels basically have the same functions as the middle and internal control levels of the DSTATCOM respectively. The control algorithm of the FESS is shown in Fig. 4.

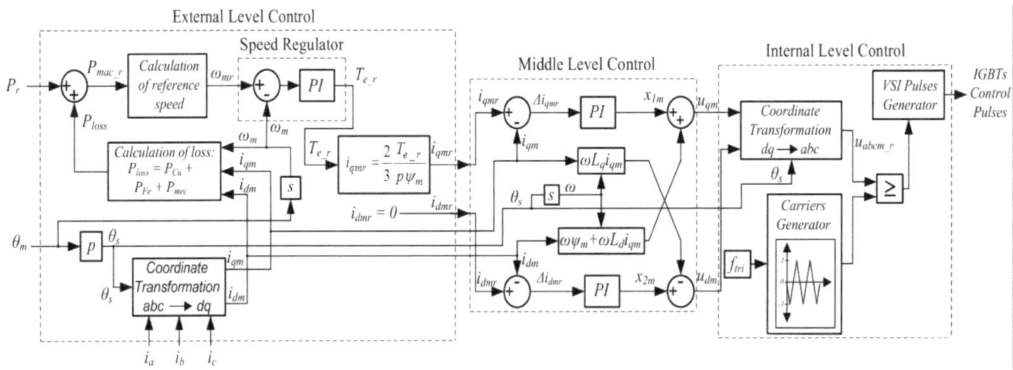

Fig. 4. Multi-level control scheme of the FESS

External level control

The external level control of the FESS is shown in simplified way on the left side of Fig. 4. In this control scheme, the reference current i_{qmr} is computed from the torque of the PMSM by using (6), and the reference current i_{dmr} is set to zero. In this way, a maximum efficiency of the PMSM is obtained (Toliyat et al., 2005).

$$T_{e\_r} = \frac{3}{2} p \psi_m i_{qmr} \tag{6}$$

where T_{e_r} is the electromagnetic torque of the machine, p the number of pairs of poles and ψ_m the magnetic flux.

The reference torque is calculated through a speed regulator which adjusts the actual speed of the machine (ω_m) to the reference speed of the machine (ω_{mr}) by using a PI controller. The reference speed is computed from the reference power of the machine, P_{mac_r}, (the power that is to be stored or delivered by the flywheel) using (7).

$$P_{mac\_r} = \frac{d\left(1/2J\omega_{mr}^2\right)}{dt} \tag{7}$$

The reference power of the machine is calculated by summing up the reference power of the DSTATCOM/FESS (P_r) and the power losses of the machine (P_{loss}). The losses of the machine are computed by summing up the copper losses (P_{Cu}) the iron losses (P_{Fe}) and the mechanical losses (P_{mec}) (Han et al., 2008).

Middle level control

A functional simplified scheme of middle level control is shown in the central part of Fig. 4. This level is basically composed of a current regulator. The control is made by using vector control; the main characteristic of this control is the synchronization of the stator flux with the rotor. The currents in the d and q axes are regulated separately. The control scheme is similar to the middle level control of the DSTATCOM, except that the synchronism angle to make the coordinate transformation, θ_s, is computed in a different way. In this case, the angle is obtained by measuring the position angle of the machine (θ_m) and multiplying by the number of pairs of poles.

Internal level control

A basic scheme of the internal level control of the FESS is shown on the right side of Fig. 4. This control level is quite similar to that of the internal level control of the DSTATCOM except that it does not have the phase locked loop block due to the fact that the angle θ_s is obtained through measurement as mentioned before.

4. Test system

The test power system used to study the dynamic performance of the DSTATCOM/FESS device proposed is shown in Fig. 5 as a single line diagram. This sub-transmission system operates at 13.8 kV/50Hz and implements a dynamically modeled wind generator linked to a bulk power system represented by an infinite bus type.

The WG (rated power: 750 kW) uses an induction generator with a squirrel-cage rotor and is connected to the grid through a transformer with star-triangle winding. The demand for reactive power from the WG is supplied by capacitors so as to reach a close-to-one power factor. The WG is modeled with blocks of an induction generator and a wind turbine available in the library of the simulation program and with parameters taken from the manufacturer data sheets (Neg Micon, 2009; Ecotècnia, 2009). The sub-transmission line is modeled by using lumped parameters. All loads are modeled by constant impedances and are grouped at bus 4 (Ld1: 0.3 MW and Ld2: 0.7 MW).

The DSTATCOM/FESS device proposed (maximum rated power: 100 kW and rated storage capacity: 750 Wh) is connected to the main bus (bus 3) through Bk 4. The DC voltage of the DSTATCOM is 750 V and the capacitor used has a rated capacitance of 1000 μF. The

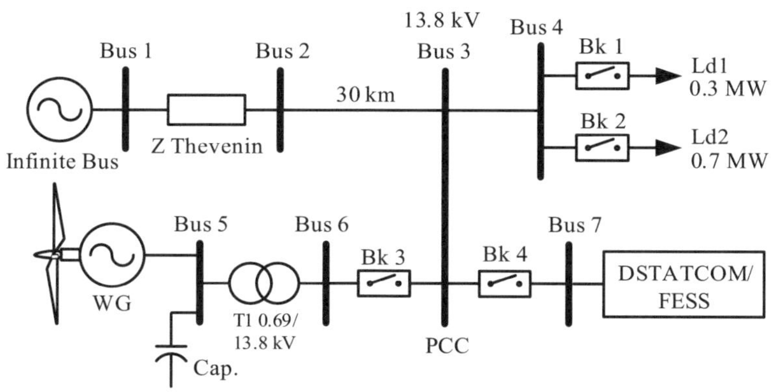

Fig. 5. Test power system

DSTATCOM-VSI works with a switching frequency of 8 kHz whereas the Interface-VSI works with 20 kHz. The parameters of the FESS (PMSM and flywheel) are obtained from the manufacturer data sheets (Beacon Power, 2009; Flywheel Energy Systems, 2009; Urenco Power Technologies, 2009).

The major test system data are summarized in Appendix A while the DSTATCOM/FESS data are in Appendix B.

The analysis and validation of the models and control algorithms proposed for the DSTATCOM/FESS controller are carried out through simple events that impose high demands upon the dynamic response of the device. A test is made of the device proposed in the test system shown in Fig. 5. For this, a variation profile of wind speed is applied to the WG so that it makes the DSTATCOM/FESS work in both ways, by storing and delivering energy. In addition, external perturbations are imposed, like a load variation, and the behaviour of the device in the different control modes is observed.

5. Simulation results

The basic system shown in Fig. 5 is used. A suitable profile for variation of the wind speed is applied, as shown in Fig. 6.

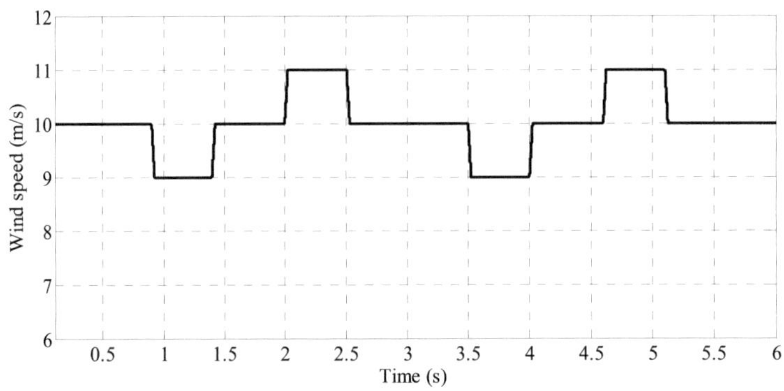

Fig. 6. Wind speed

The wind speed variations cause significant fluctuations in the active and reactive power injected by the WG. The capacitor bank used with the WG is adjusted to compensate for the reactive power when the WG operates at a mean wind speed of 10 m/s. In bus 4, the load Ld1 = 0.3 MW is first connected (in t = 0 s) and then, in t = 3 s, the load Ld2 = 0.7 MW is added. The behaviour of the system is analyzed in both cases, when the DSTATCOM/FESS is disconnected (Bk 4 opened) and connected (Bk 4 closed). The variations of active power injected by the WG-DSTATCOM/FESS system for both cases are shown in Fig. 7. With the DSTATCOM/FESS device connected, the variations of power from the WG are reduced and an active power that is practically constant is injected to the system.

For the reactive power control, three different cases are presented: DSTATCOM/FESS disconnected, DSTATCOM/FESS connected working in Power Factor Control Mode (PFCM); and DSTATCOM/FESS connected working in Voltage Control Mode (VCM).

The reactive power injected by the WG-DSTATCOM/FESS system is shown in Fig. 8. With the DSTATCOM/FESS connected working in PFCM, it is observed that the reactive power injected by the WG-DSTATCOM/FESS system is zero. Consequently, the device proposed has satisfactorily compensated for the reactive power variations of the WG. With the DSTATCOM/FESS connected working in VCM, the reactive power variations from the WG are compensated for and the device also generates or consumes the reactive power necessary to make the voltage in bus 4 be 1 pu.

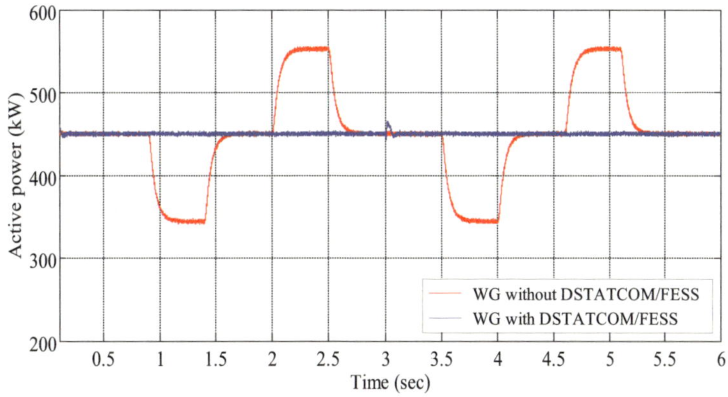

Fig. 7. Active power of the WG-DSTATCOM/FESS system

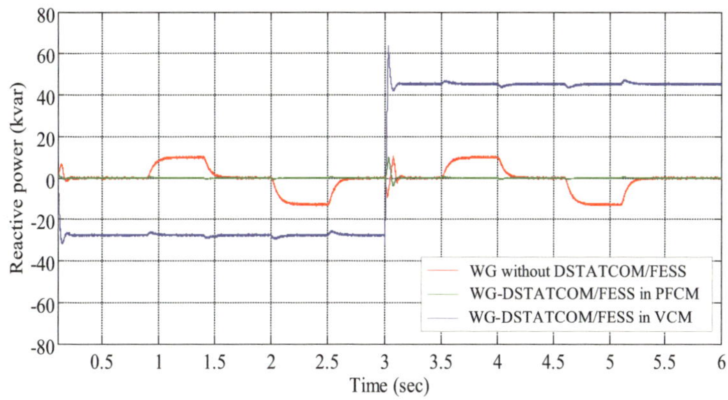

Fig. 8. Reactive power of the WG-DSTATCOM/FESS system

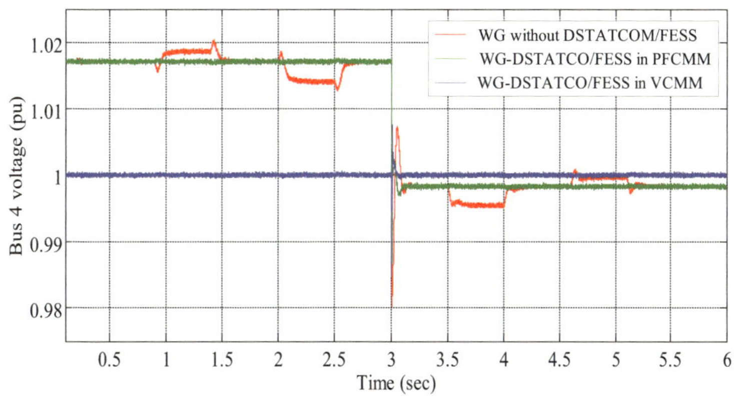

Fig. 9. Voltage at bus 4

The voltage at bus 4 is shown in Fig. 9. When there is no compensation, the voltage has significant variations due to both power variations from the WG and those of the load. When the DSTATCOM/FESS device is connected in PFCM, there are no voltage variations due to wind power variations. However, this mode has the problem that the voltage has a value different from 1 pu and it varies with disturbances in the load. When the DSTATCOM/FESS device is connected in VCM, the voltage is maintained at 1 pu independently of the variations in wind power and variations of the load. This control mode solves in quite effective way the problem observed with the PFCM. Therefore, the VCM is the most convenient mode when the connection point of the WG does not have any other device that dynamically controls the voltage.
The active and reactive (in PFCM and VCM) power generated by the DSTATCOM/FESS is shown in Fig. 10. The electromagnetic torque and the rotational speed of the PMSM are shown in Fig. 11 and Fig. 12 respectively. It can be observed that when the storage device works by delivering energy (positive active power), the PMSM has a positive torque applied, making the speed goes down and thereby releasing stored kinetic energy. Then, when the storage device works by storing energy (negative active power), the PMSM has a negative torque applied, making the speed goes up with a consequent increase in the stored kinetic energy.

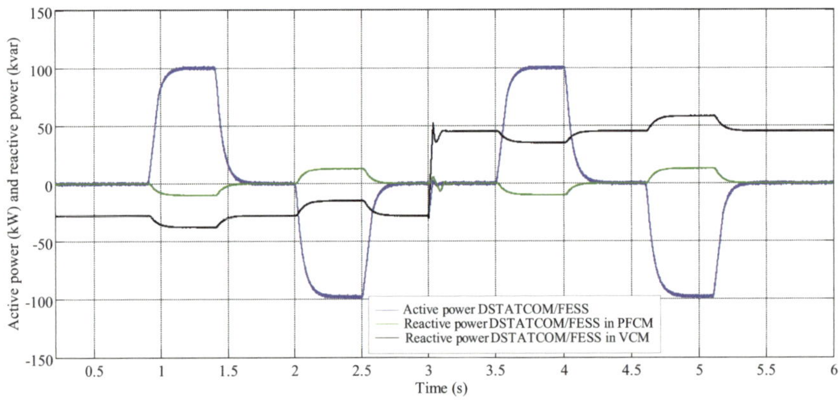

Fig. 10. Input/output active and reactive power of the DSTATCOM/FESS

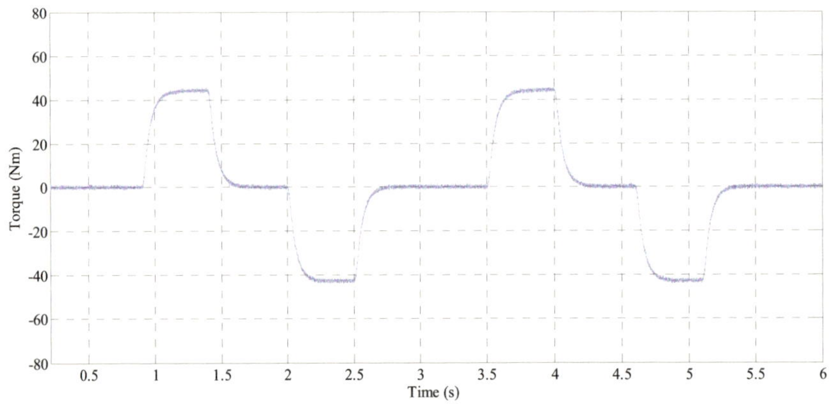

Fig. 11. Electromagnetic torque of the PMSM

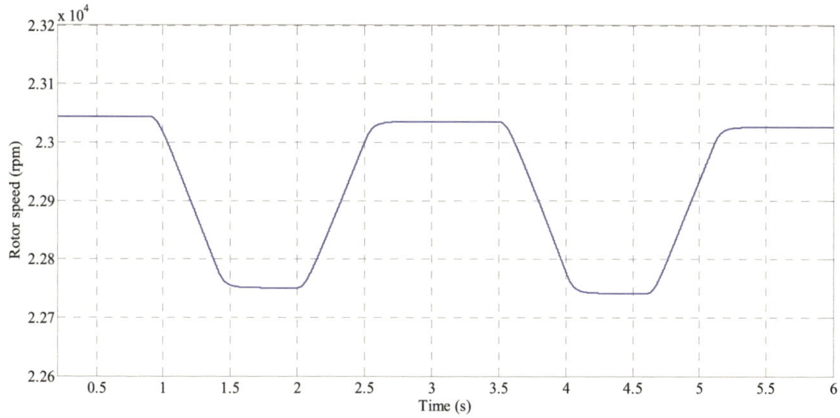

Fig. 12. Rotor speed of the PMSM

The i_d and i_q currents of the DSTATCOM measured at the low voltage side of the transformer are shown in Fig. 13. An excellent decoupling can be observed among the currents comparing Fig. 13 with Fig. 10. Variations of the i_d current correspond with those of the active power and variations of the i_q current correspond with those of the reactive power. In addition, the i_d and i_q currents of the PMSM are shown in Fig. 14. It is worth noting that under the power requirements imposed, only a variation of i_q exists. In this way, the condition imposed on the control to keep i_d equal to zero so as to obtain a maximum efficiency of the PMSM holds true.

The currents of the DSTATCOM and the PMSM in the abc reference frame are shown in Fig. 15 and Fig. 16 respectively. The currents shown belong to an interval of the maximum power transfer. It can be observed a high frequency of the current of the PMSM (around 800 Hz). This is due to the high rotational speed of the PMSM.

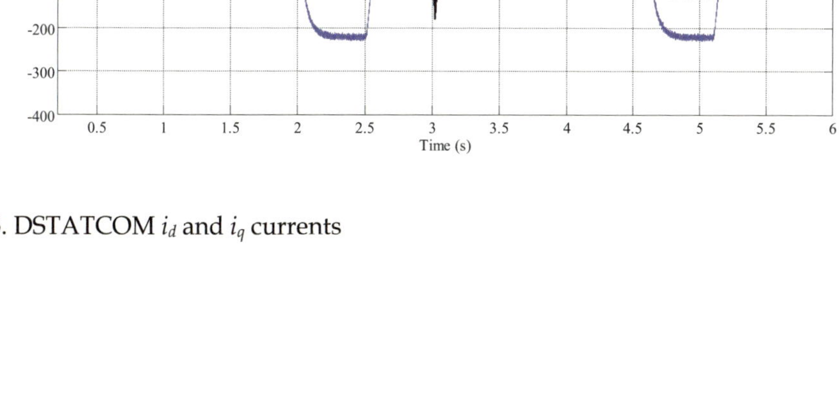

Fig. 13. DSTATCOM i_d and i_q currents

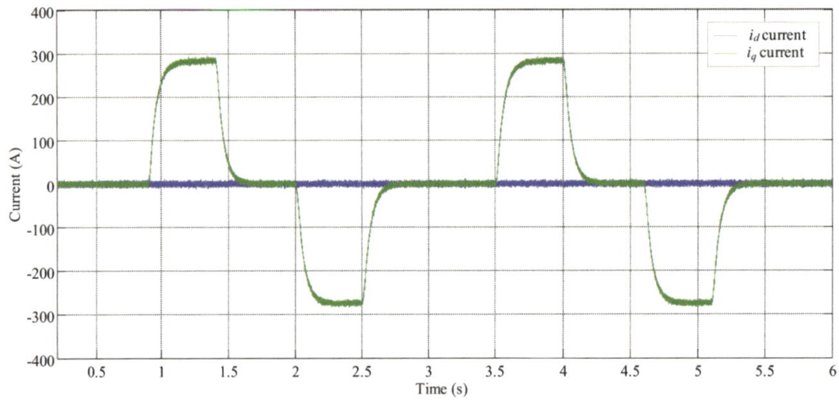

Fig. 14. PMSM i_d and i_q currents

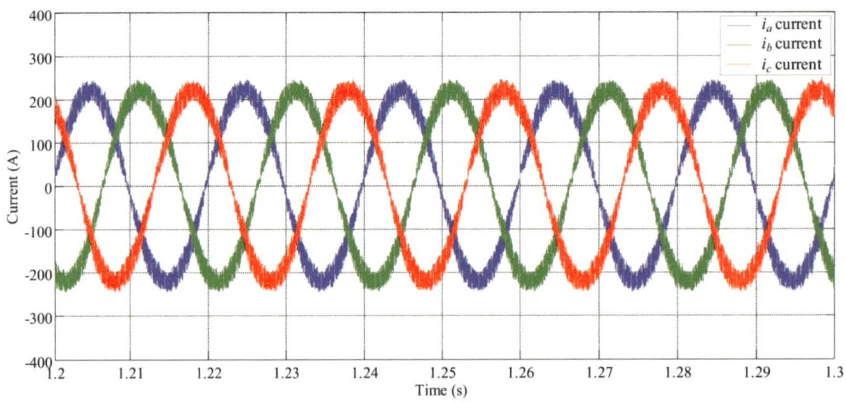

Fig. 15. DSTATCOM i_a, i_b and i_c currents

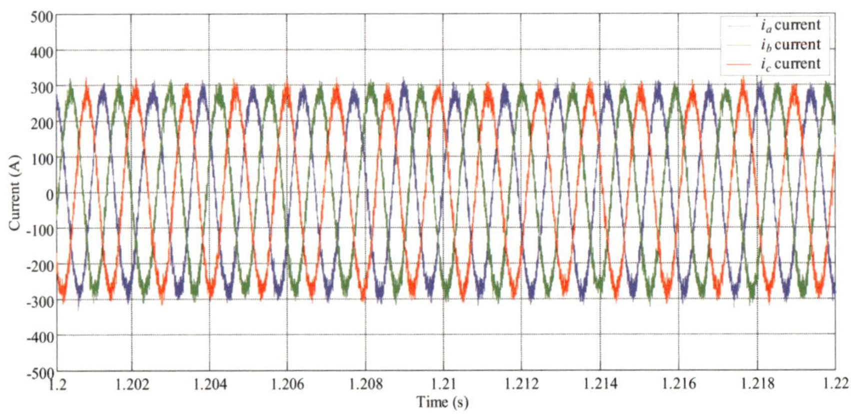

Fig. 16. PMSM i_a, i_b and i_c currents

Efficiency and losses of the PMSM

Tests were made for different power requirements in the whole range of operation speeds of the machine. The results of the efficiency of the machine for an exchange of power of 10 kW, 50 kW and 100 kW are shown in Fig. 17. In this figure a high efficiency of the PMSM can be observed above 98% when the power is high (50 or 100 % of the rated power), even in the whole speed range. A division of the losses of the PMSM for different requirements of power are shown in Fig. 18. In these figures it can be observed that the mechanical and the iron losses increase with the rotational speed of the machine and they practically do not depend on the exchange power. Moreover, it can be observed that the copper losses depend both, on the rotational speed and on the exchange power. The copper losses have more significant values at low speeds. This is because in order to deliver a certain constant power at low speeds a bigger torque and therefore a bigger current are required. When there is no power transfer, the losses range from 0.3-1.1 kW.

6. Conclusions

This paper presents model aspects and control algorithms of a DSTATCOM controller coupled with a High-Speed Flywheel Energy Storage System. A proposal is made of a detailed fully realistic model of the compensator and a novel multi-level control algorithm taking into account three control modes to mitigate problems introduced by wind power in power systems.

From the results obtained, it can be concluded that the detailed models and developed control algorithms have worked satisfactorily. With the implemented control, an excellent decoupling is kept in the control of the active and reactive power. Moreover, with the device and control modes proposed, the power fluctuations coming from a WG are effectively compensated. It was shown that the WG-DSTATCOM/FESS system can deliver a constant active power in a time range of seconds or more, depending on the storage capacity. For the reactive power control, it was shown that the system proposed is able to provide a unitary power factor or to obtain a dynamic control of the voltage in the connection point for power disturbances in the WG and also for fluctuations in the system such as sudden variations in the load. Therefore, the incorporation of DSTATCOM/FESS has shown that it can improve the power quality in wind power systems.

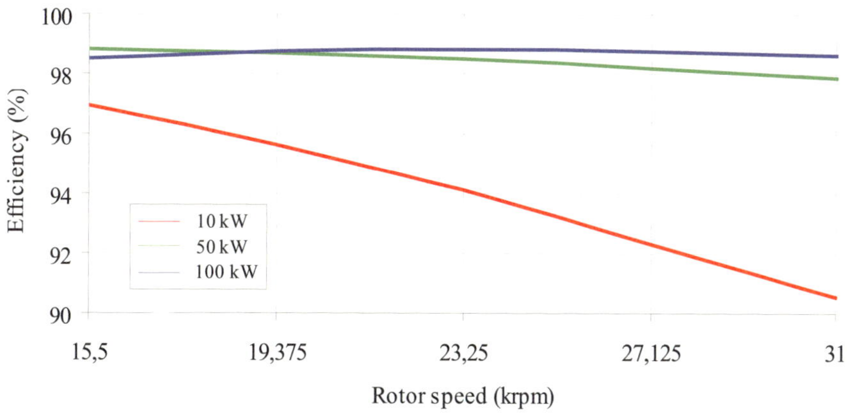

Fig. 17. Efficiency of the PMSM

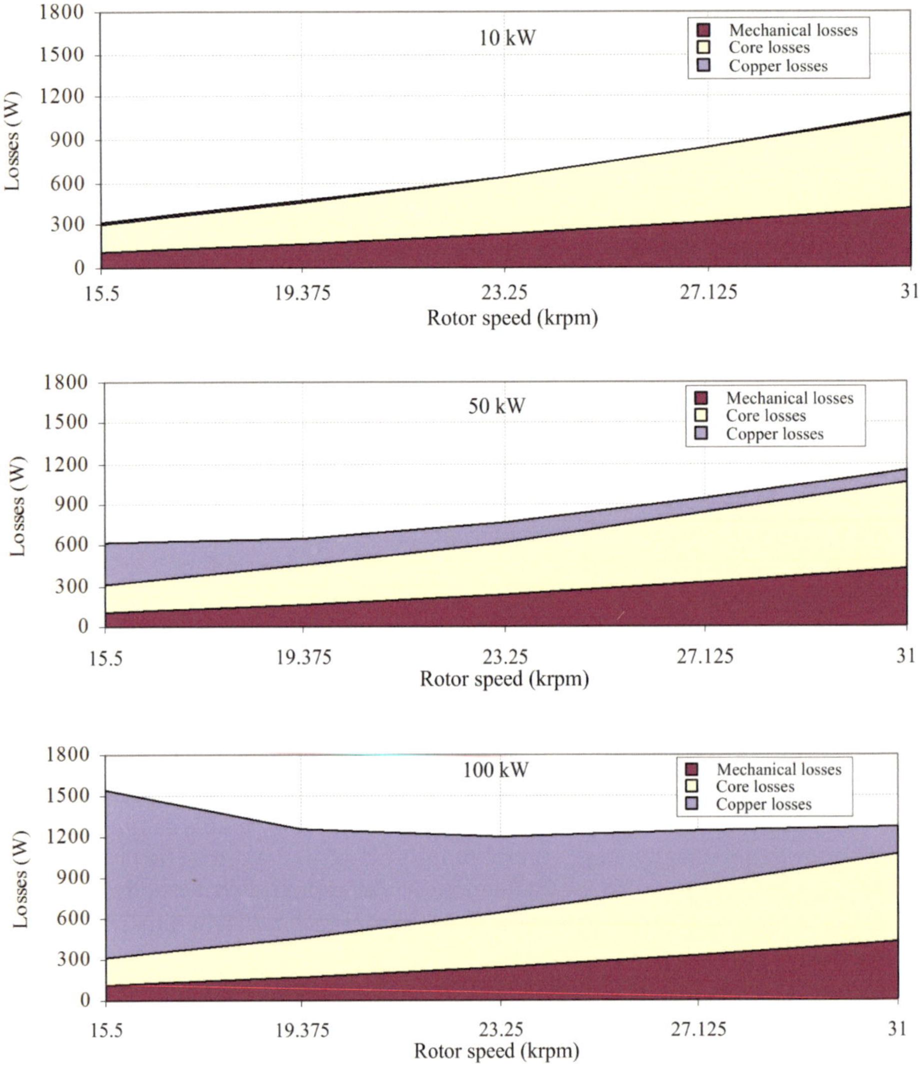

Fig. 18. Losses of the PMSM for transfer power of: 10kW, 50kW and 100kW

7. APPENDIX A

TEST SYSTEM DATA

Line data are given in Table 1. Table 2 shows the transformer data. All p.u. quantities are on 13.8 kV and the transformer rated MVA base. Table 3 shows the main parameters of the generation unit coupled to the wind turbine. Table 4 shows the main parameters of the wind turbine and the power curve of the turbine is shown in Fig. 19. All p.u. quantities are on a 690 V and on the 750 kVA base. Finally, the most important load data are shown in Table 5.

ID	From bus	To bus	U_N kV	L km	R Ω/km	X Ω/km	B μΩ^{-1}/km
L1	2	3	13.8	30	0.01273	0.2933	4.0024

Table 1. Line data
ID: component identifier; U_N: rated voltage; L: line length; R, X and B: positive sequence resistance, reactance and susceptance of sub-transmission line.

ID	From bus	To bus	R pu	X pu	Rm pu	Xm pu	S_N kVA	N_p/N_s kV/kV
T1	5	6	0.002	0.021	500	500	1000	0.69/13.8

Table 2. Transformer data
R and X: winding resistance and reactance; Rm and Xm: magnetization resistance and reactance; S_N: rated power; N_p/N_s: voltage transformation ratio

ID	Bus	Machine	Rotor	S_N kVA	U_N V	Rs pu	Xs pu	Rr pu	Xr pu	H s	p
WG	5	Induction	Squirrel-cage	750	690	0.016	0.06	0.016	0.06	0.095	2

Table 3. Wind generator data
Rs and Xs: stator resistance and reactance; Rr and Xr: rotor resistance and reactance; H: inertia constant; p: pairs of poles

ID	H s	$Wc\text{-}i$ m/s	$Wc\text{-}o$ m/s	Wrp m/s
WT	2	4	25	16

Table 4. Wind turbine data
Wc-i: cut-in wind speed; *Wc-o*: cut-out wind speed; *Wrp*: rated wind speed

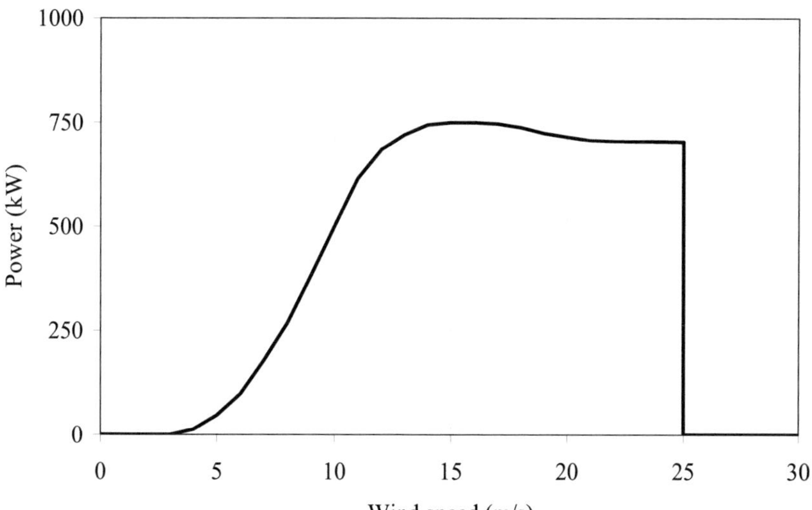

Fig. 19. Power curve of the wind turbine

ID	Bus	P_L kW	Q_L kvar
Ld1	4	300	0
Ld2	4	700	0

Table 5. Load data
P_L and Q_L: load real and reactive power.

8. Appendix B

DSTATCOM/FESS controller data

Tables 6-8 summarize the most important data corresponding to the FESS, Interface and DSTATCOM subsystems.

General							
ID	P_{max} kW	E Wh	t_d s	S_{min} krpm	S_{max} krpm	J kg m^2	U_d V
FW	100	750	27	15.5	31	0.72	750

Table 6. FESS data
P_{max}: maximum rated real power; E: rated storage capacity; t_d: discharge time; S_{min} and S_{max}: minimum and maximum operation speed; J: Polar inertia (PMSM + flywheel); U_d: DC voltage.

PMSM				
Motor/ Generator	ψ_m Wb	L_d, L_q μH	R mΩ	p
Permanent Magnet 3-phase, synchromous	0.052	100	8	2

Table 7. PMSM data
ψ_m: flux induced by magnet; L_d and L_q: d and q axis inductances; R: resistance of the stator windings.

T_f μs	T_t μs	U_f V	R_{on} mΩ	R_s kΩ
1	2	1	1	100

Table 7. VSI data of the Interface and the DSTATCOM
T_f: Current 10 % fall time of the IGBT, T_t: Current tail time of the IGBT; U_f: forward voltage for IGBTs; R_{on}: internal resistance of the IGBT device; R_s: snubber resistance

9. References

Ackermann, T. (2005). *Wind Power in Power systems*. John Wiley & Sons, Ltd, ISBN 0-470-85508-8 (HB), England.

Andrade, R.; Sotelo, G. G.; Ferreira, A. C.; Rolim, L. G. B.; da Silva Neto, J. L.; Stephan, R. M.; Suemitsu, W. I. & Nicolsky, R. (2007). Flywheel Energy Storage System Description and Tests, IEEE Transactions on Applied Superconductivity, Vol. 17, N° 2, (June 2007), ISSN: 1051-8223.

Barton, J. P. & Infield, D. G. (2004). Energy storage and its use with intermittent renewable energy, *IEEE Transaction Energy Conversion*, Vol. 19, N° 2, pp. 441–448, (June 2004), ISSN: 0885-8969.

Beacon Power website, www.beaconpower.com/, May 2009.

Bose, B. K. (2002). *Modern Power Electronics and AC Drives*, Prentice Hall - 2002, ISBN 0-13-016743-6, United States of America.

Boutot, T.; Chang, L. & Luke, D. (2002). A Low Speed Flywheel System for Wind Energy Conversion, *Proceedings of the 2002 IEEE Canadian Conference on Electrical & Computer Engineering*, 0-7803-7514-9/02, Winnipeg, May 2002, Canada.

Brad, R. & McDowall, J. (2005). Commercial Successes in Power Storage. *IEEE power & energy magazine*, Vol. 3, No. 2, (March/April 2005) pp. 24-30, ISSN 1540-7977.

Cárdenas, R.; Peña, R.; Asher, G. M.; Clare, J. & Blasco-Giménez, R. (2004). Control Strategies for Power Smoothing Using a Flywheel Driven by a Sensorless Vector-Controlled Induction Machine Operating in a Wide Speed Range, *IEEE Transactions on Industrial Electronics*, Vol. 51, No. 3, (June 2004) 603-614, ISSN: 0278-0046.

Carrasco, J. M. (2006). Power Electronic System for Grid Integration of Renewable Energy Source: A Survey, *IEEE Transaction on Industrial Electronics*, Vol. 53, No. 4, pp 1002-1014, (August 2006), ISSN : 0278-0046.

Cimuca, G.; Radulescu, M.M.; Saudemont, C. & Robyns, B. (2004). Comparative Study of Flywheel Energy Storage Systems Associated to Wind Generators, *Proceedings of the International Conference on Applied and Theoretical Electricity* - ICATE 2004, Oct 2004, Romania.

Chen, Z. & Spooner, E. (2001). Grid Power Quality with Variable Speed Wind Turbines. *IEEE Transactions on Energy Conversion*, vol. 16, N° 2, pp 148-154, June 2001.

Ecotècnia website, www.ecotecnia.com, March 2009.

Flywheel Energy Systems website, www.magma.ca/~fesi, May 2009.

Han, S.; Jahns, T.M. & Zhu, Z. Q. (2008). Analysis of Rotor Core Eddy-Current Losses in Interior Permanent Magnet Synchronous Machines, *IEEE, Industry Applications Society Annual Meeting*, IAS '08, October 2008.

Hebner, R. ; Beno, J. & Walls, A. (2002). Flywheel batteries come around again, *IEEE Spectrum*, Vol. 39, No. 4, pp. 46–51, (April 2002), ISSN: 0018-9235.

Mohod, S.W. & Aware, M.V. (2008). Power Quality Issues & It's Mitigation Technique in Wind Energy Generation. *IEEE Harmonics and Quality of Power*, September 2008.

Molina M. G. & Mercado, P. E. (2004). Multilevel control of a Static Synchronous Compensator combined with a SMES coil for applications on Primary Frequency Control, *Proc. CBA 2004*, Gramado, Brasil, Septiembre 2004.

Neg Micon website, www.neg-micon.com, March 2009.

Samineni, S.; Johnson, B. K.; Hess, H. L. & Law, J. D. (2006). Modeling and Analysis of a Flywheel Energy Storage System for Voltage Sag Correction, *IEEE Transactions on Industry Applications*, Vol. 42, No 1, (Janaury/February 2006), 1813-1818, ISSN: 0093-9994.

Slootweg, J.G. & Kling, W.L. (2003). Is the Answer Blowing in the Wind? *IEEE Power & Energy magazine*, pp 26-33, November/December 2003.

Smith, J.C.; Milligan, M.R. & DeMeo, E.A. (2007). Utility Wind Integration and Operating Impact State of the Art. *IEEE Transaction on Power System*, vol. 32, N°.3, pp.900-907, August 2007.

Song, Y. H. & Johns, A. T. (1999). *Flexible AC Transmission Systems (FACTS)*, IEE Press, ISBN 0-85296-771-3. London, UK.

Suvire, G. O. & Mercado, P. E. (2007). Utilización de Almacenadores de Energía para Mitigar los Problemas Introducidos por la Generación Eólica en el Sistema Eléctrico, *Décimo Segundo Encuentro Regional Ibero-americano del CIGRÉ*, Foz do Iguazú-Pr, Brasil, Mayo 2007.

Suvire, G. O. & Mercado P. E. (2008). Wind Farm: Dynamic Model and Impact on a Weak Power System, *IEEE PES* T&D *LATINAMERICA*, pp. 1-8, ISBN: 978-1-4244-2217-3, Bogotá-Colombia, August 2008.

Takahashi, R.; Wu, L.; Murata, T., & Tamura, J. (2005) An Application of Flywheel Energy Storage System for Wind Energy Conversion, *International Conference on Power Electronics and Drives Systems*, Vol. 2, pp 932-937, 2005.

Toliyat, H.; Talebi, S.; McMullen, P.; Huynh C. & Filatov A. (2005). Advanced High-Speed Flywheel Energy Storage Systems for Pulsed Power Applications, IEEE Electric Ship Technologies Symposium, 2005.

Urenco Power Technologies website, http://uptenergy.com, May 2009.

Xie, H.; Mei, S. & Lu, Q. (2002). Design of a Multi-Level Controller for FACTS Devices, *Proc. Power Systems and Communication Infrastructures for the Future*, Pekín, China, September 2002.

3

The High-speed Flywheel Energy Storage System

Stanisław Piróg, Marcin Baszyński and Tomasz Siostrzonek
University of Science and Technology
Poland

1. Introduction

At the present level of technology the electricity generation has already ceased to be a problem. However, years are passing by under the slogan of seeking for methods of effective energy storage. The energy storage method shall be feasible and environmentally safe. That's why the methods, once regarded as inefficient, are recently taken into consideration. The development in materials technology (carbon fibre, semiconductors, etc.) brought back the concept of a flywheel. This idea has been applied to high-speed flywheel energy storage.

2. Electromechanical energy storage using a flywheel

A flywheel energy storage system converts electrical energy supplied from DC or three-phase AC power source into kinetic energy of a spinning mass or converts kinetic energy of a spinning mass into electrical energy.
The moment of inertia of a hollow cylinder with outer radius r_z, and inner radius r_w is:

$$J = \frac{1}{2}\pi h \rho \left(r_z^4 - r_w^4 \right) \tag{1}$$

Maximum amount of kinetic energy stored in a rotating mass:

$$W_{k\max} = \frac{1}{2} J \omega_{\max}^2 = \frac{\pi}{4} h \rho \left(r_z^4 - r_w^4 \right) \omega_{\max}^2 \tag{2}$$

where: J - moment of inertia, ω - angular velocity.
The force acting on a segment of spinning hoop (Fig. 1) is:

$$dF_r = dm \cdot \frac{v^2}{r} = \rho \cdot h \cdot d\varphi \cdot dr \cdot v \tag{3}$$

where: ρ - density of the hoop material, h - height, r - radius, v - peripheral velocity, φ - angle, F - force, m - mass.
The net force acting in the direction of axis x, resulting from elementary forces dF_r, is:

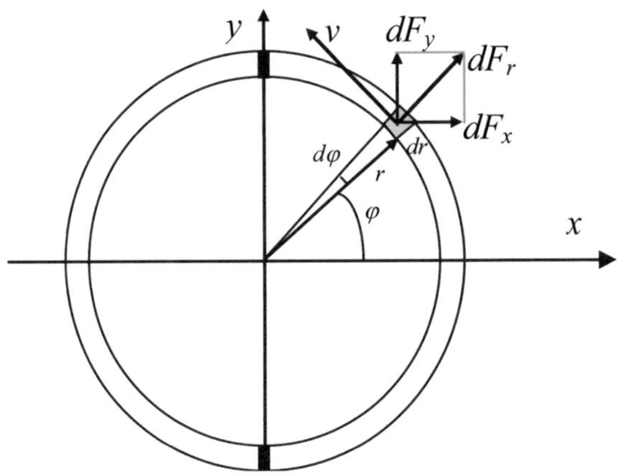

Fig. 1. Forces acting on the segment of a rotating hoop

$$F_x = 2\int_0^{\frac{\pi}{2}} dF_r \cos\varphi \cdot d\varphi = 2\rho \cdot h \cdot dr \cdot v^2 \int_0^{\frac{\pi}{2}} \cos\varphi \cdot d\varphi = 2\rho \cdot h \cdot dr \cdot v \tag{4}$$

Bursting stress (in the hoop cross sections shaded in Fig. 1):

$$\sigma_r = \frac{F_x}{2 \cdot h \cdot dr} = \frac{2\rho \cdot h \cdot dr \cdot v^2}{2 \cdot h \cdot dr} = \rho \cdot v^2 \tag{5}$$

Hence, the maximum allowable peripheral velocity for a material with the density ρ and allowable tensile stress $R_e = \sigma_{r\max}$:

$$v_{\max}^2 = \frac{R_e}{\rho} \tag{6}$$

Maximum rotational velocity of a flywheel depends on the allowable peripheral velocity at its surface (6):

$$\omega_{\max}^2 = \frac{v_{\max}^2}{r_z^2} = \frac{R_e}{r_z^2 \rho} \tag{7}$$

Substituting (7) into (2) we have:

$$W_{k\max} = \frac{\pi}{4} h\rho\left(r_z^4 - r_w^4\right)\frac{R_e}{r_z^2 \rho} = \pi\left(r_z^2 - r_w^2\right)h\frac{\left(r_z^2 + r_w^2\right)R_e}{r_z^2 \rho} = V\left(1 + \left(\frac{r_w}{r_z}\right)^2\right) \tag{8}$$

Hence can be found the flywheel mass:

$$m = \pi h\left(r_z^2 - r_w^2\right)\rho = \frac{4W_{k\max}}{R_e} \cdot \frac{1}{1 + \left(\frac{r_w}{r_z}\right)^2}\rho \tag{9}$$

In order to minimize the flywheel mass it shall be made in the form of a thin-walled hollow cylinder.
From relation (9) the ratio of maximum stored energy to the flywheel mass is:

$$\frac{W_{k\max}}{m} = \frac{W_{k\max}}{\rho V} = \frac{R_e}{\rho} \cdot \frac{1 + \left(\frac{r_w}{r_z}\right)^2}{4} \tag{10}$$

For $r_z \approx r_w$ relation (10) reduces to the form of:

$$\frac{W_{k\max}}{m} \approx \frac{R_e}{2\rho} = \frac{v_{\max}^2}{2} \tag{11}$$

As follows from (11), a light structure (a large amount of energy per unit of mass) can be achieved using a material with possible low density ρ and high tensile strength R_e. Materials that meet these requirements are composites (Kevlar, carbon fibre, glass fibre in combination with a filler) or composite bandage (in order to improve stiffness) on a ring of a light metal, e.g. aluminium.

	Density ρ [kg/m³]	Strength Re [GPa]	v_{max} [m/s]	W/m [MJ/kg]
Steel	$7.8 \cdot 10^3$	1.8	480.4	0.23
Titanium	$4.5 \cdot 10^3$	1.2	516	0.27
Composite glass fibre	$2.0 \cdot 10^3$	1.6	894.4	0.80
Composite carbon fibre	$1.5 \cdot 10^3$	2.4	1256	1.60

Table 1. Parameters of typical flywheel materials

A flywheel of a larger energy per unit of mass and the given outer radius r_z, chosen for constructional reasons, has to rotate with a higher peripheral velocity (11) and, consequently, with a higher angular velocity (7).
Since in this case peripheral velocities of high-speed rotors are exceeding the speed of sound, the rotor should be enclosed in a hermetic vacuum chamber. In consequence, the energy store structure - and particularly bearings, become complicated (due to vacuum maintained in inside the enclosure should be used magnetic bearings and a system stabilizing the rotor axle position in space The flywheel, integrated with the electric machine, should rotate without a contact with motionless parts (magnetic levitation). Magnetic bearings should be made of permanent magnets (high efficiency is required) while an electromagnetic system should only assist them to a certain extent and stabilize the axle position. Due to a required very high efficiency, the flywheel shall be driven by a permanent magnet motor installed inside the enclosure. Vacuum inside the enclosure prevents exchange of heat between the FES components and causes problems with heat removal from windings of the electric machine operated as a motor or generator. An advantage of vacuum is lack of losses caused by the rotor friction in air (at peripheral velocities of 700-1000m/s) and noiseless operation.

The electric machine must be controlled by a power electronic system enabling its operation as a motor or generator and adjusting electric power parameters alternately to the needs of the accelerated spinning mass or electrical loads (or an electric network) supplied from FES. If the energy storage system is operated as an autonomous energy source (isolated operation) it must be provided with a power electronic system that prohibits propagation of load unbalance (the output voltage double-frequency ripple component) to the flywheel torque.

The amount of energy stored in FES is proportional to the square of angular velocity. It means that at the 1/3 of maximum velocity remains only ca. 10% of maximum energy. The energy store should be therefore operated within the speed range from 1/3 to maximum speed. The voltage at the electric machine winding changes with the ratio 1:3, and the power electronic system shall be designed to tolerate such changes. in order to minimize losses (conduction losses in semiconductor devices) the maximum voltage applied to electric machine should be possibly high (up to 1000V).

The design of an energy storage system that meets up-to-date requirements is an interdisciplinary and complex engineering task that requires the use of the-state-of-the art technologies and materials. The energy storage system can be applied to:

- Power quality improvement systems to compensate active power peaks and limit their impact on power supply network and reduce peak loads. Required are: a large stored energy (of the order of hundreds MJ) and large instantaneous power that enables discharging during a tens of seconds.
- Standby power supplies to backup or start other power sources (a motor-generator set or switching to another network) for particularly important and sensitive processes.
- Systems for storage and controlled release of energy produced by alternative autonomous electric power sources, like photovoltaic or wind power plants. In such systems store energy in time when there is no demand from electricity users. A flywheel energy storage system intended for supporting alternative autonomous sources shall exhibit very high energy efficiency (due to the necessity of long accumulation time) and three-phase output with possibility for unbalanced load at constant frequency (50 Hz) and constant rms voltage magnitude. The amount of stored energy is ca. 5÷10 MJ.
- Limiting wind farms power fluctuations by means of a dynamic accumulation of peak power generated during high-wind periods and release it during low-wind periods.
- Accumulation (storage) of energy recovered from regenerative braking of intermittently started and stopped (or reversing) large-power drives (e.g. rolling mills and winders) or energy recovered from discharging large electromagnets.
- Elevators in buildings with intensive traffic flow ("intelligent building"). An elevator equipped with an energy storage system will consume energy solely to compensate losses.
- Large industrial plants (large-power flywheel energy storage systems) in order to mitigate voltage fluctuations, power supply back-up during supply systems switching, and power quality improvement by means of peak loading and unloading reduction. Reduction of peak active power will result in reduced transmission losses and enable the use of more economical installations (smaller cross-sectional areas, transformer powers, etc.), smaller peak contracted power.
- Urban buses. Flywheel energy storage systems designed for mobile applications with relatively small energy stored (6÷10 MJ) and suitable for charging and discharging with large powers (100÷150 kW) can be utilized in urban buses (charged at bus stops).

- Urban and suburban electric transportation systems and hybrid vehicles (internal combustion engine, generator, electric motor), flywheel energy storage systems can absorb kinetic energy of a braking vehicle and reuse it during travel.

3. Technical requirements for flywheel energy storage systems

- High efficiency.
- Small mass and volume.
- Reliability, durability and safety.
- Capability for operation in a three-phase power network or autonomous operation with unbalanced load.
- Large short-duration power (capability for quick charging and discharging).

4. Electric machine for the flywheel energy storage purposes

Flywheel energy storage systems can utilize all types of AC three-phase machines. The choice of the machine type is determine by the energy storage application and particularly by expected duration of energy storage. In energy storage systems with expected long duration of energy storage idle losses should be radically limited. Idle losses in systems with long duration of energy storage should be radically limited. Such systems can utilize asynchronous induction machines or synchronous machines. During energy charging or discharging a small amount of energy is needed for the machine excitation (power losses in the field winding resistance in a synchronous machine or losses due to the magnetizing (reactive) component in an induction machine). In energy storage systems intended for relatively short duration storage, permanent magnet machines (synchronous or brushless) can be used. In flywheel energy storage systems with a high rotational speed and, consequently, high frequency of the fundamental component of the machine voltage, the difficulty lies in correct shaping of sinusoidal current waveform obtained by means of PWM modulation. In such a case a correct power supply of a brushless DC machine can be more easily achieved. Permanent magnet machines require no additional energy for excitation but certain small losses occur in them due to currents induced in conducting parts by variable magnetic field of rotating magnets. These losses can be reduced employing brushless coreless machines. Such machines have very small winding inductance and in order to achieve a continuous current they require additional external reactors when supplied from PWM modulated inverters.

5. Examples of flywheel energy storage applications

In an autonomous system with alternative electric energy source (Fig. 2a) the energy store supplies loads if loss of supply from a base power source occurs. The energy storage can be used in uninterruptible power supply systems (UPSs) of selected loads (Fig. 2b). Upon voltage loss or decrease in the line voltage magnitude a load and energy storage system are instantaneously disconnected (by means of thyristor switches) from the supply line and energy store turns to the generator mode, thereby powering sensitive (critical) loads Another application of an energy storage system is stabilization of supply voltage (or limitation of peak currents in a supply line) of loads characterized with fast-changing, short-duration loading far exceeding the average load.

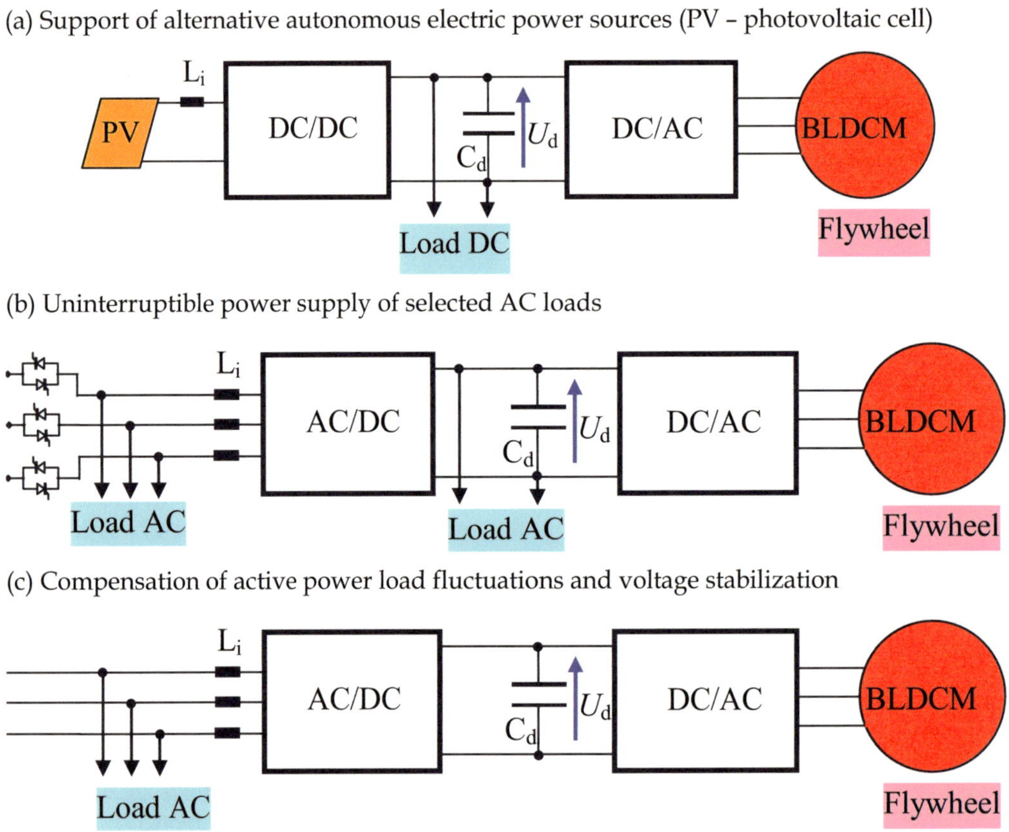

Fig. 2. Examples of spinning energy storage applications; AC/DC, DC/AC - power electronic converters, BLDCPM - electric machine (Brushless D.C. Permanent Magnet Motor)

6. Controlling energy release from a flywheel energy storage system

The amount of energy stored in a rotating mass is proportional to the angular velocity squared. It means that energy store can be effectively utilized within the range from maximum angular velocity (W_{max}) to 1/3 of angular velocity (1/9 W_{max}). There are several solutions for limiting the maximum power of energy release from (or supplied to) the energy store.

Figure 3 shows the relative energy (W/W_m) and power (P/P_m) vs. relative angular velocity (ω/ω_m). Line (1) is the characteristic of a storage system operated within the velocity range $(0.5 \div 1)\omega_{\max}$ with limited power. The consequence of the power limitation is the necessity for limiting the current maximum value according to relation $I_{d\max} = P_{\max} / U_m$ (curve 3). Line (2) represents the power change for operation with the current maximum value determined by the straight line (4).

Another control method consists in operation with constant maximum power within the angular velocity range $(0.5 \div 1)\omega_{\max}$. Characteristics of the storage system controlled employing this method is shown in Fig. 4. A boundary of the control method can be set at a lower velocity; this results in limiting maximum power to a lower value.

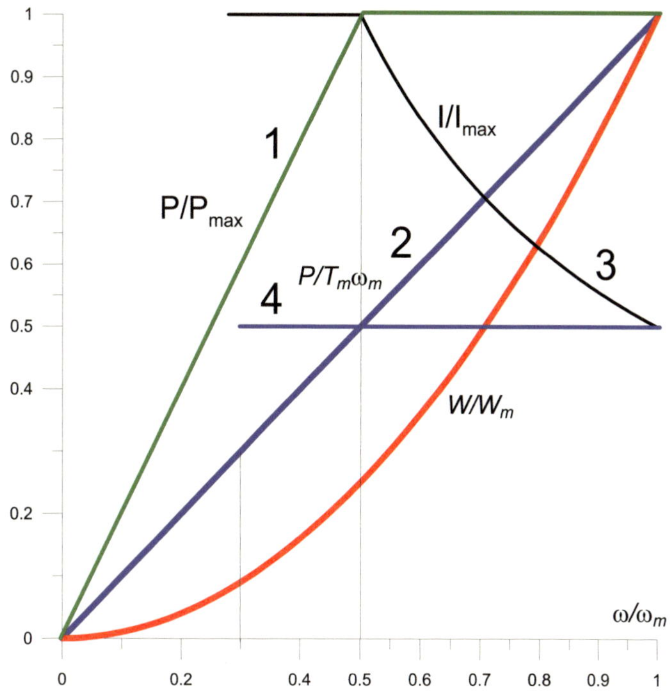

Fig. 3. Characteristic of a energy storage system with $P_{max}=f(\omega)$

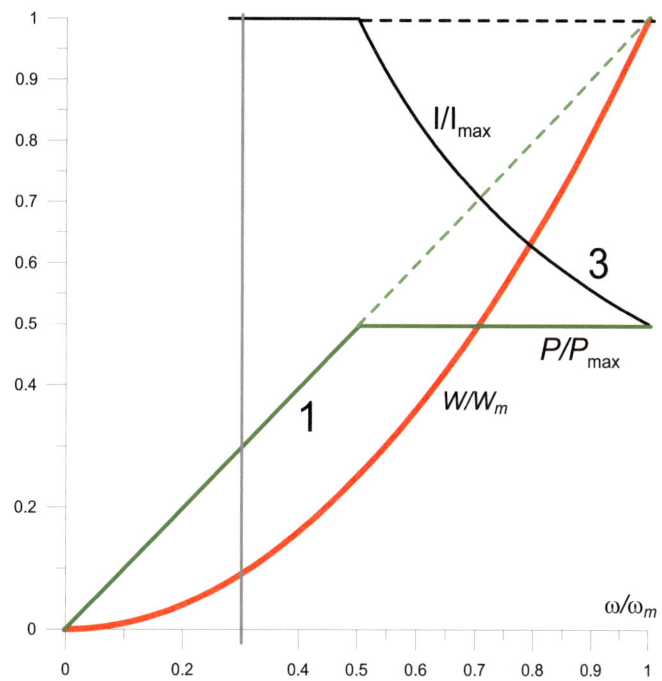

Fig. 4. Characteristic of a energy storage system with $P/P_{max}=const$ for $\omega>\omega_{max}/2$

It should be borne in mind that energy of 1kWh (3.6 MJ) is equivalent to potential energy of the mass 1000 kg at the height of 367 m, i.e. the release the amount of energy (equivalent to that consumed by a 100 W bulb during 10 hours) required to throw a 1-ton car to the height of 367 m ($3.6 \cdot 10^6$[J] =1000[kg]·9.81[m/s^2]h hence h =367[m]; air friction and the car and ground deformations are not taken into account).

7. Permanent magnet motors

Permanent magnet motors combine features of classical DC separately excited motors with advantages of an induction motor drive. They are manufactured in many structural variations with respect to both the permanent magnets arrangement and the method of their fixing, as well as the motor applications (permanent magnets in the stator or rotor). In terms of the current and the back electromotive force waveforms, permanent magnet machines can be categorized into two types:

- Permanent Magnet Synchronous Motor (PMSM),
- Brushless Permanent Magnet DC Motors (BLDCM, BLDC, BLPMDCM).

Permanent magnet synchronous motors (*PMSM*) exhibit properties similar to those of synchronous AC machines. They are characterized by:

- sinusoidal distribution of magnetic flux in the air gap,
- sinusoidal phase currents,
- sinusoidal back electromotive force (*BEMF*).

In a brushless permanent machine the back electromotive force has a trapezoidal waveform and the required current waveform has the form of rectangular, alternating sign pulses. Idealized relations between the back electromotive force and phase currents are shown in figure 5.

In order to provide a constant torque the machine should be supplied in such a manner that the instantaneous power value remains constant (in figure 5 the instantaneous power waveform in each phase is indicated green). This requirement is met for rectangular phase currents Duration of both the positive and negative pulse is T/3, time-interval between pulses is T/6, and phase-shift between phases is T/3. During each time interval T/6 the current is conducted simultaneously only in two phases. The motor instantaneous power is the sum of powers generated in two phases. The electromagnetic torque is the quotient of the instantaneous power and the motor angular velocity. At constant angular velocity the torque is constant only if the instantaneous power is constant.

A brushless DC permanent magnet motor cannot, as a machine, be supplied without supplementary equipment, thus its integral components are:

- a power electronic converter that provides power supply of appropriate phase windings depending on the rotor position,
- a controller stabilizing the current depending on the required torque (Fig. 6).

8. Bipolar PWM of an inverter supplying a brushless DC permanent magnet motor

The pulse-width modulated voltage-source inverter, supplying a brushless DC permanent magnet motor enables shaping the required phase currents waveform by means of the supply voltage control.

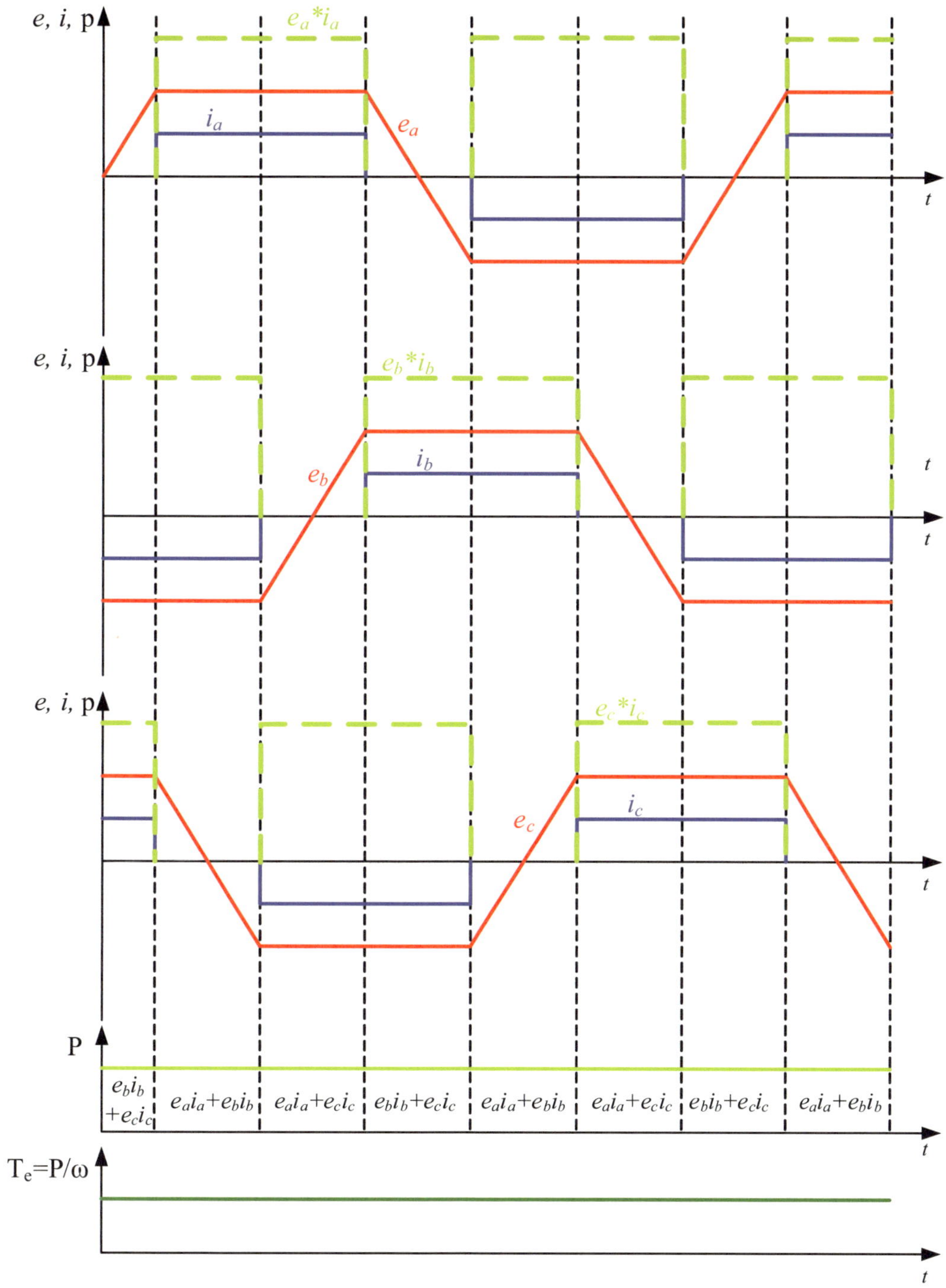

Fig. 5. Desired waveforms of electromotive force, phase currents, instantaneous power and electromagnetic torque

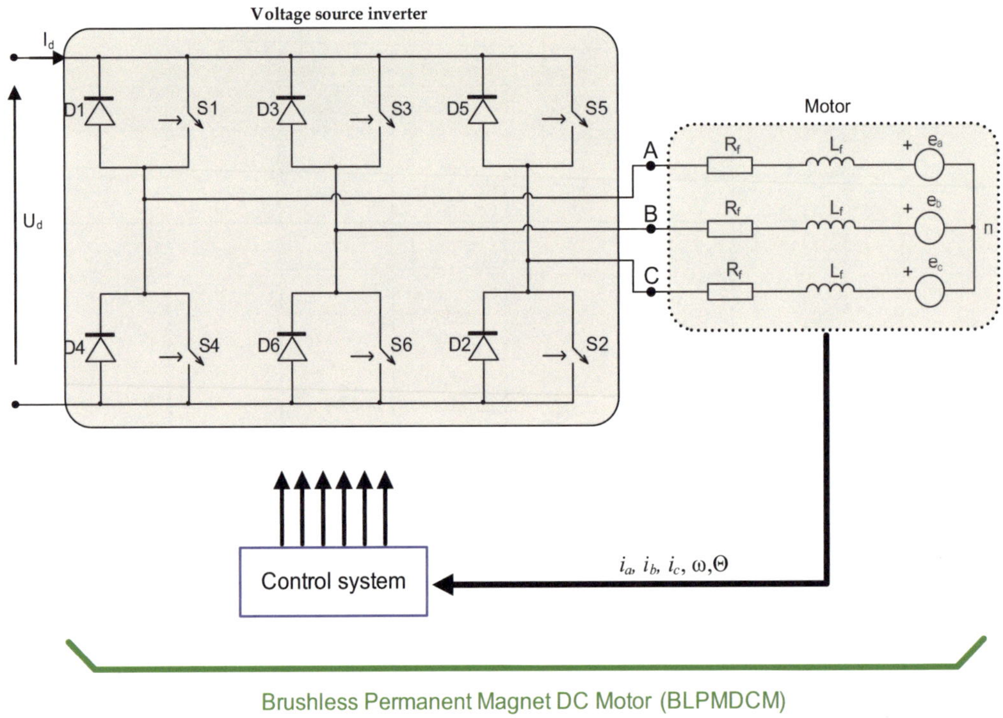

Fig. 6. A brushless permanent magnet DC motor supplied from a voltage source inverter with control system

Where this type of control is employed, only two switches are chopper controlled during the time interval of duration T/6. The sequence of switching is shown in figure 7. The inverter is controlled in the same manner as a single-phase inverter. The switches pairs, e.g. S1 and S6, are switched during the time interval equal T/6. The current flows through two phases A and B connected in series. After elapse of time equal T/6 switch S6 stops conducting and switch S2 is turned on to conduct (chopper controlled) together with the switch S1. Phase A is still connected to the DC voltage source positive terminal, phase B is being connected to its negative terminal. The current flows in phases A and C connected in series. Switch S1 is active during time period T/3 During each time interval with duration T/6 one of the phases is disconnected from both terminals of the DC voltage source, switches are switched specifically at T/6 intervals. At each time-instant the converter operates as a single-phase inverter and can be analysed as such. The inverter configurations with individual switches turned on are shown in figure 8.

9. Torque control of brushless permanent magnet DC machine

Figure 9 shows phase currents (i_a, i_b, i_c), their modules ($|i_a|$, $|i_b|$, $|i_c|$), the sum of the modules ($\Sigma|i|$) and torque (T_e). Apart from the fast-changing torque component resulting from finite time of semiconductor devices PWM switching, also torque ripple occurs due to the current commutation between the motor phase windings. Thus in each 1/6 of the period a noticeable disturbance occurs in the torque waveform.

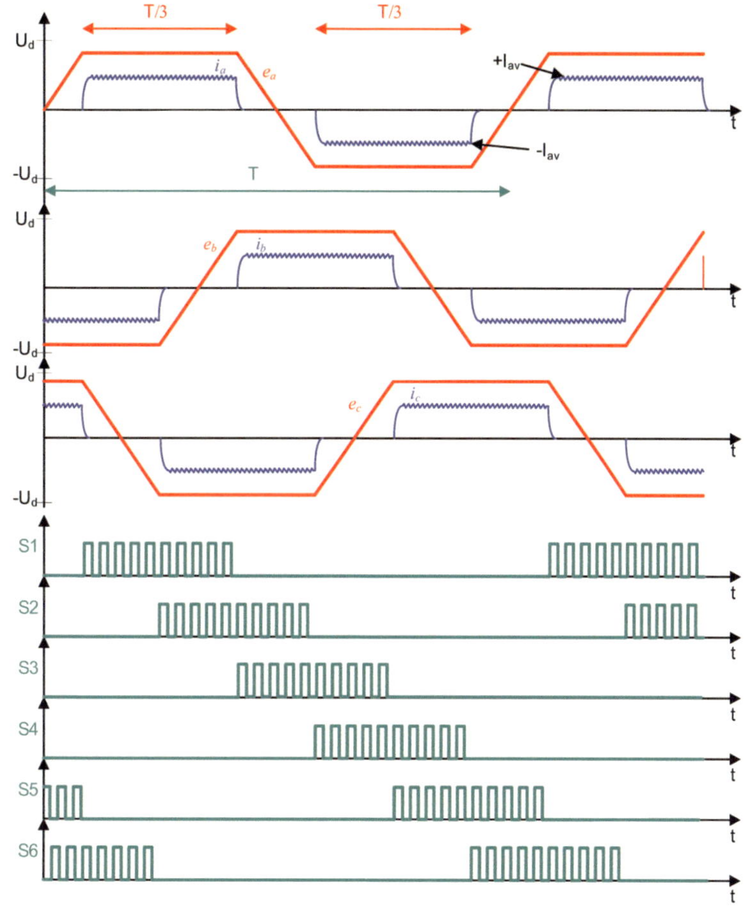

Fig. 7. Bipolar pulse width modulation: phase currents and switch control pulses

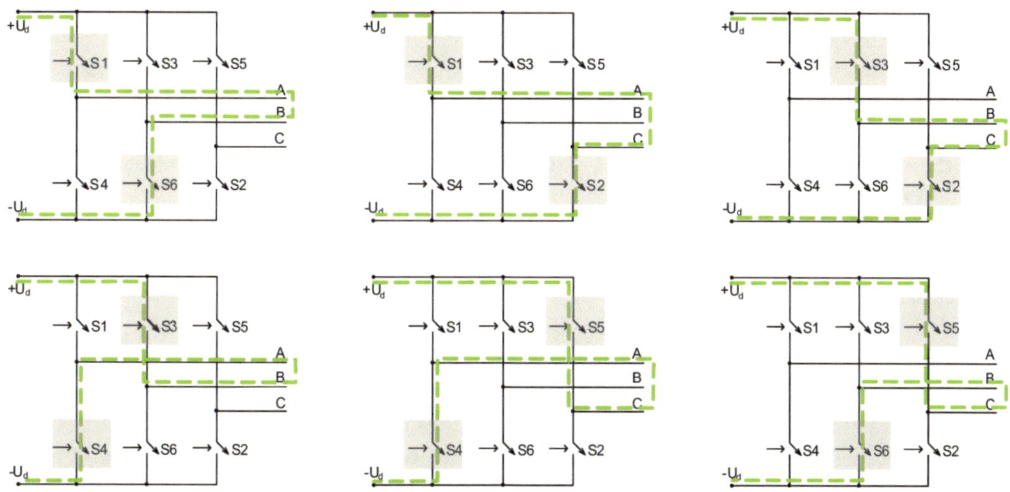

Fig. 8. Bipolar pulse width modulation: the sequence of switching

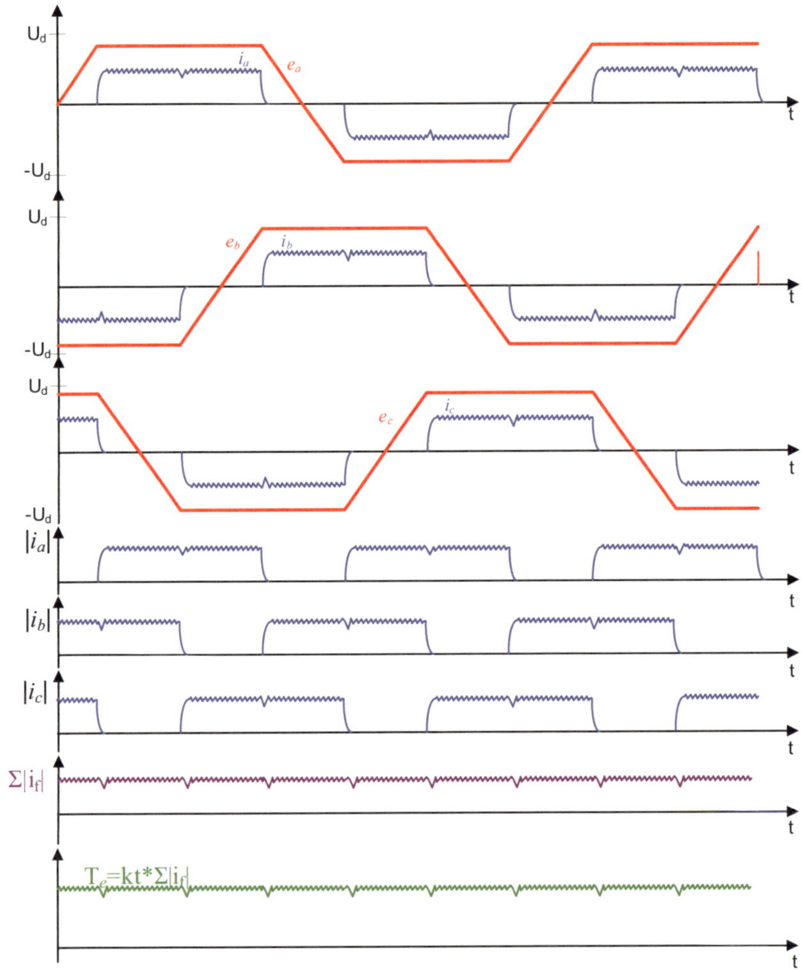

Fig. 9. Actual waveforms of phase currents, their modules and electromagnetic torque

The brushless machine torque is controlled by means of the phase currents control. The control is achieved, similarly as in a classical shunt DC machine, by modulation of fixed frequency pulses width by the output signal of a PI current controller. The feedback signal should be proportional to the actual value of the DC source current module. It can be obtained in two ways:

- measuring the module of the converter input current (DC source current) (Fig. 10), or
- measuring phase currents; the feedback signal is proportional to the sum of the load rectified phase currents (Fig. 11).

A drawback of the first solution is an additional inductance (of the sensor and its connections) connected between the capacitor and semiconductor devices. The inverter should be supplied from a voltage source and the incorporated inductance changes the source character during transient states. This inductance is the source of overvoltages occurring across semiconductor devices that require overvoltage protection in the form of RC snubber circuits to absorb overvoltage energy. These additional components increase both the system complexity and power losses in the converter.

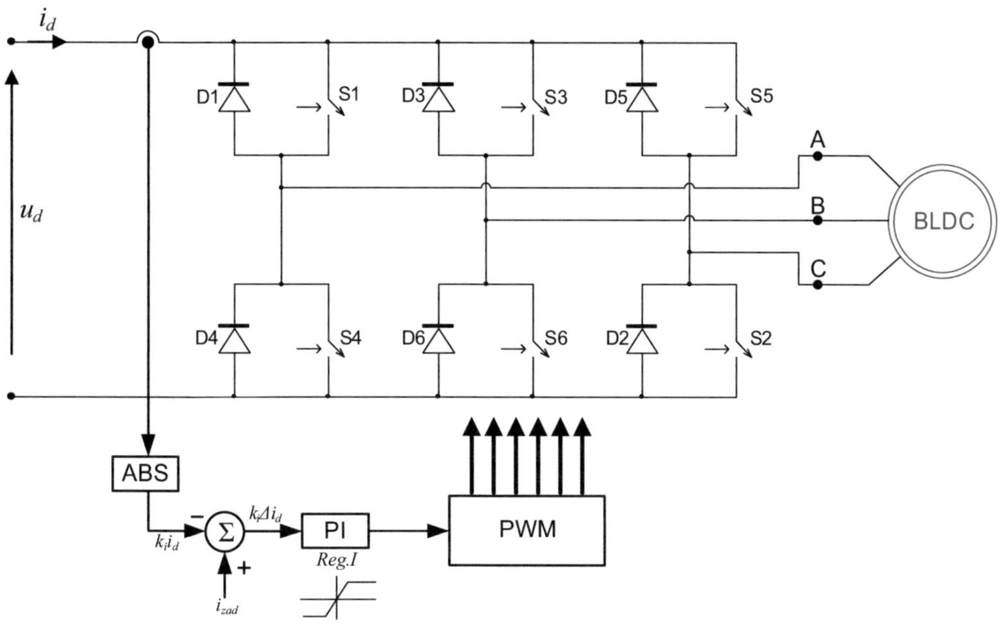

Fig. 10. Measurement of the inverter input current

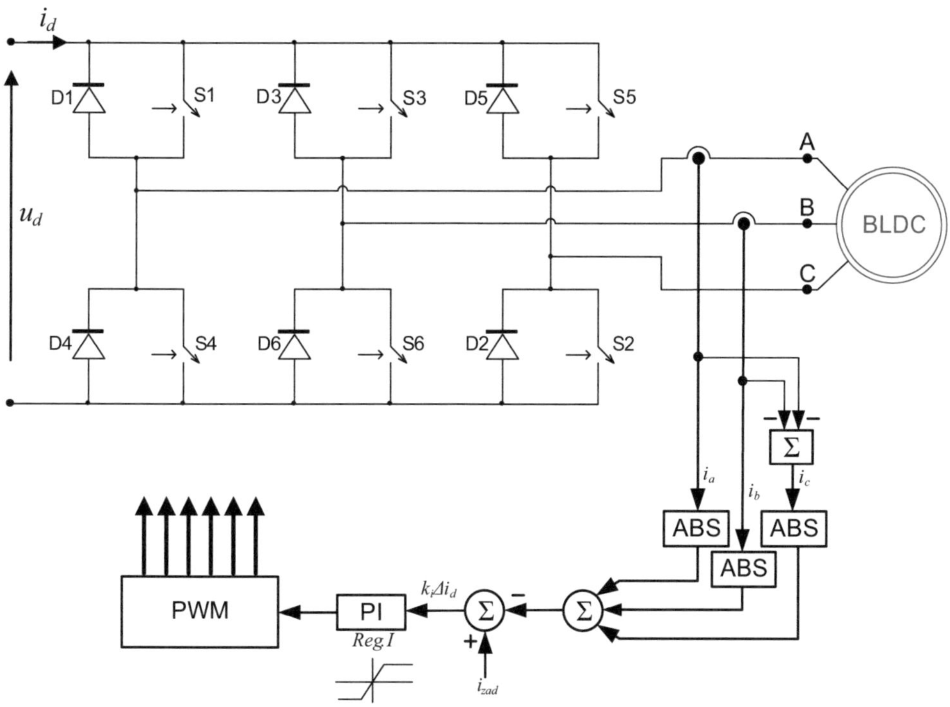

Fig. 11. The feedback signal circuit utilizing the phase currents measurement

Apart from current components from controlled switches, also the currents of backward diodes occur in the DC source current. These currents, flowing in the direction opposite to the switches current, result from the magnetic field energy stored in the machine windings and transferred back to the DC source. The phase current value depends on both these components. Therefore, in order to obtain the feedback signal, the absolute value of the signal proportional to the measured DC source current has to be taken.

The second way the feedback signal can be obtained is the measurement of phase currents. Since $i_a+i_b+i_c=0$ it is sufficient to use transducers in the load two phases. The signal proportional to the DC source current is obtained by summing the absolute values of phase currents (Fig. 11). The error signal is the difference between the DC current reference and the actual source current, reconstructed from the measured phase currents. In the pulse width modulation a high-frequency triangle carrier signal is compared with the current controller output signal. The current controller output signal limit is proportional to the phase-to-phase peak voltage value. That way are generated control pulses of fixed frequency and modulated width to control the inverter transistors switching.

10. Determining the rotor poles position relative to stator windings

Figure 12 shows the cross section of a brushless permanent magnet DC motor. The motor is assumed to have a single pole-pair rotor while the stator winding has three pole-pairs. Figure 13 shows waveforms of the current and back electromotive force in phase A depending on the mutual positions of characteristic points. The analysis starts at the instant when point K coincides with point z_1. At his time the magnet N-pole begins overlapping the stator pole denoted by a. The back electromotive force (BEMF) increases linearly until the stator pole is completely overlapped by the magnet N-pole. This takes T/6. Then, the magnetic flux increases linearly during T/3 thus the back electromotive force is constant. The rectangular waveform of the current in phase A is shaped by means of chopper control.

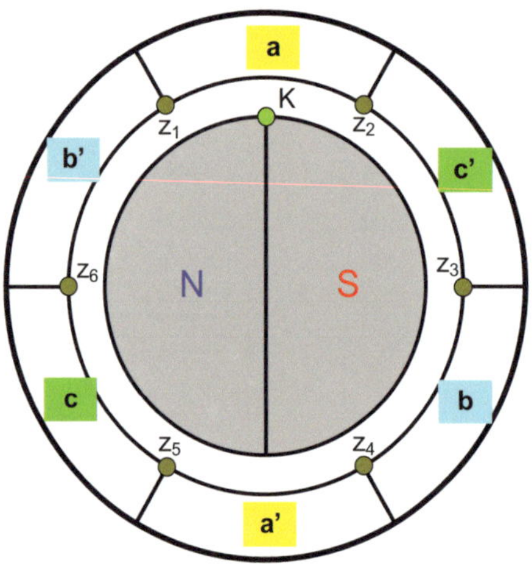

Fig. 12. The cross section of a BLDCM motor

Since point K coincides with z_4 the back electromotive force decreases linearly until point K is in the position where N-pole begins overlapping the stator pole denoted a'. Between the point z_5 and z_1 the back electromotive force is constant and negative.

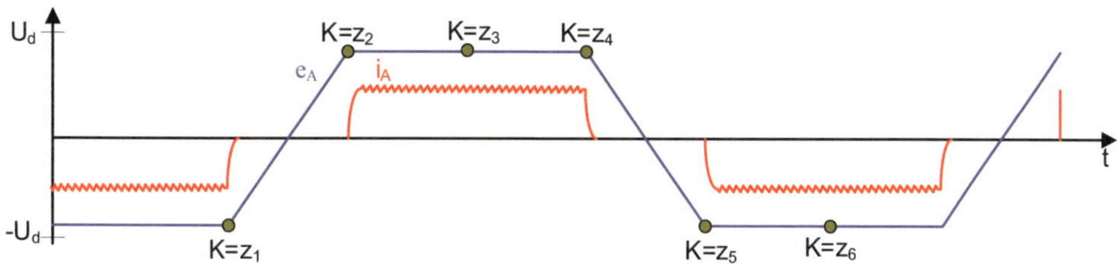

Fig. 13. Waveforms of the current and back electromotive force in one phase depending on the permanent magnet poles position

In motors with trapezoidal BEMF it is essential that voltage switching on or off to a given winding is synchronized with the rotor position relative to this winding axis.

11. AC/DC converter

A unity input power factor control of a three-phase step-up converter is feasible in the rotating co-ordinate frame because in this system the source frequency quantities are represented by constant values. The diagram of the rectifier connection to a supply network is shown in figure 1. Since X_L>>R, the resistances of reactors are disregarded in the diagram.

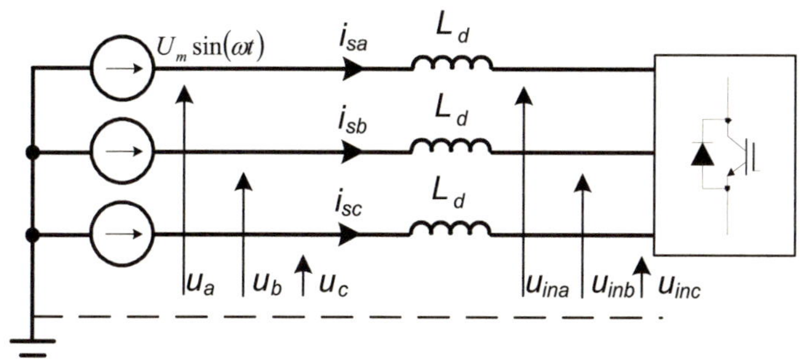

Fig. 14. Diagram of the rectifier connection to a supply network

The following designations are used the diagram of figure 1: i_{sn}– phase currents, u_{sn}– the supply line phase-to-neutral voltages, u_{inn}– the converter output voltage (where $n= a, b, c$). The phase currents, according to the diagram, are described by equation (12).

$$u_{sn} - u_{inn} = L\frac{di_{sn}}{dt} \tag{12}$$

Converting the equation (12) into the rotating reference frame dq we obtain equation (13).

$$\boldsymbol{u}_{sdq} - \boldsymbol{u}_{indq} = \Delta\boldsymbol{u}_{dq} = L_d\frac{d\boldsymbol{i}_{sdq}}{dt} + j\omega L_d\boldsymbol{i}_{sdq} \tag{13}$$

Decomposing the equation (12) into dq components we obtain (14).

$$u_{ind} = u_{sd} - \Delta u_d = u_{sd} - (L_d\frac{di_{sd}}{dt} - \omega L_d i_{sq}) \tag{14}$$

$$u_{inq} = u_{sq} - \Delta u_q = u_{sq} - (L_d \frac{di_{sq}}{dt} - \omega L_d i_{sd}) \tag{15}$$

Equations (14) and (15) describe the converter input voltages. Inserting the required line current values into these equations we can determine the output voltage waveforms forcing the required current. The components $L_d(di_{sdq}/dt)$ represent the converter dynamic states (load switching or changes in the load parameters). Assuming the control system comprises only proportional terms we obtain from equations (14) and (15) relationships describing the control system (16) and (17).

$$u_{ind} = u_{sd} - (K_R \Delta i_{sd} - K_d \Delta i_{sq}) = u_{sd} - [K_R(i_{sdr} - i_{sd}) - K_d(i_{sqr} - i_{sq})] \tag{16}$$

$$u_{inq} = u_{sq} - (K_R \Delta i_{sq} - K_q \Delta i_{sd}) = u_{sq} - [K_R(i_{sqr} - i_{sq}) - K_q(i_{sdr} - i_{sd})] \tag{17}$$

Figure 15 shows block diagram of the control system and the power circuit. The following designations are used in the diagram: *TP* – switch-on delay units (blanking time),

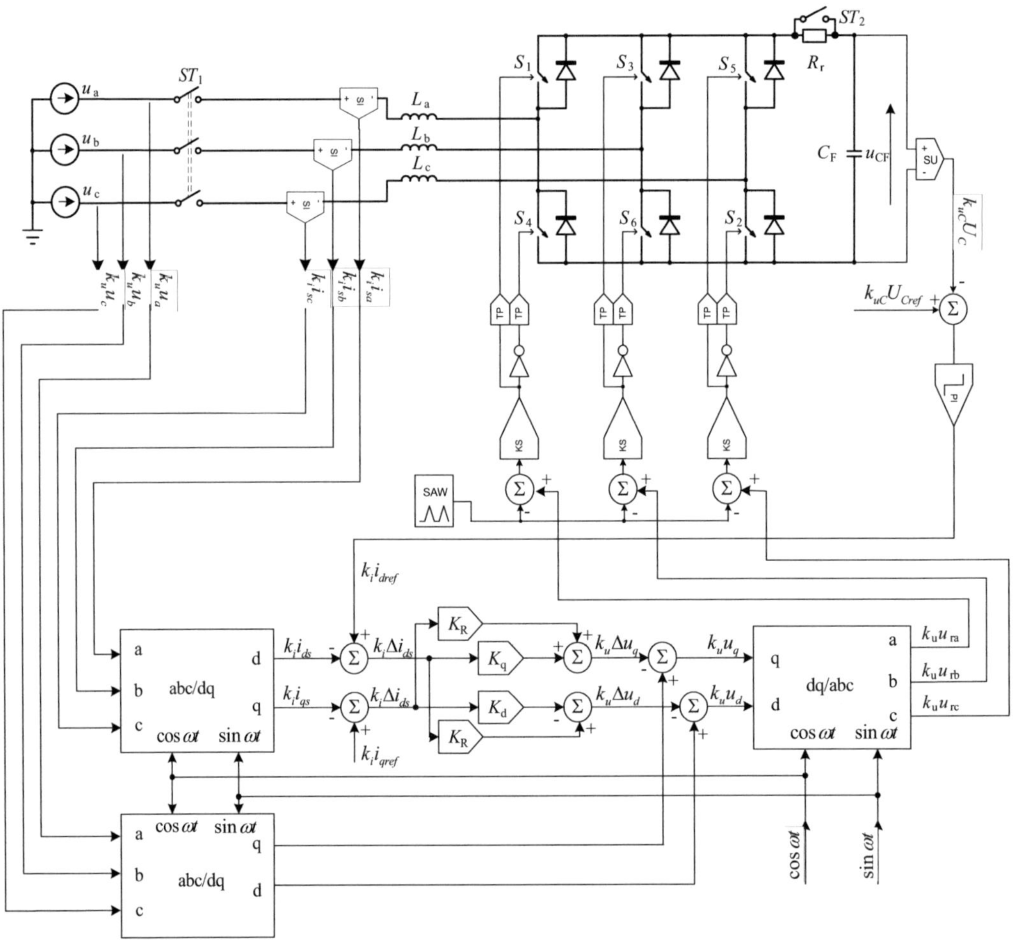

Fig. 15. Block diagram of the control system and the power circuit

PI - proportional-integral controller, *KS*- sign comparator, *SAW*- triangle wave generator, K_R, K_d, K_q- proportional terms, *ST*- contactors, R_a, R_b, R_c - resistors limiting the capacitor charging current, Σ- adder.
The control circuit of diagram 15 employs transformation from the thee-phase system to the rotating co-ordinate system (abc→dq), described by equation (12).

$$\begin{bmatrix} v_d \\ v_q \end{bmatrix} = \begin{bmatrix} \cos\omega t & \sin\omega t \\ -\sin\omega t & \cos\omega t \end{bmatrix} \begin{bmatrix} v_a \\ \frac{1}{\sqrt{3}}(v_b - v_c) \end{bmatrix} \tag{18}$$

Where:

$$\begin{cases} v_a = V_m \cos\omega t \\ v_b = V_m \cos(\omega t - \frac{2}{3}\pi) \\ v_c = V_m \cos(\omega t + \frac{2}{3}\pi) \end{cases} \tag{19}$$

11.1 Synchronization circuit

In order to determine the transformation abc→dq it is necessary to generate functions $\cos\omega t$ and $\sin\omega t$, as follows from equation (18), such that the function $\cos\omega t$ will correspond (i.e. be cophasal) to $v_a = V_m \cos\omega t$. In practical solutions various methods for generating the $\cos\omega t$ and $\sin\omega t$ functions are employed, e.g. synchronization with a single, selected phase (normally *a*) employing a single-phase PLL loop. The advantage of this method is an easy implementation in digital technique. Microprocessor systems employ an external, specialized device performing the functions of a phase-locked loop, connected with a microprocessor port dedicated for counting external events. Therefore the CPU workload due to generating the $\cos\omega t$ and $\sin\omega t$ functions is reduced to minimum. A drawback of this method is the generated function is related to only one phase of the synchronizing signal and the system does not control the other phases. In the event of a disturbance starting in phase *c* (a phase jump in the synchronizing voltage caused by switching a large active power load) the control system will respond with large delay. In order to protect the converter from effects of a phase jump the synchronization circuit should control all phases of the synchronizing voltage. Substituting equations (19) describing the three-phase synchronizing voltage into equation (18), the transformation abc→dq takes the form (20).

$$\begin{bmatrix} v_d \\ v_q \end{bmatrix} = \begin{bmatrix} V_m(\cos^2\omega t + \sin^2\omega t) \\ V_m(-\sin\omega t\cos\omega t + \sin\omega t\cos\omega t) \end{bmatrix} = \begin{bmatrix} V_m \\ 0 \end{bmatrix} \tag{20}$$

It follows from equation (20) that if the functions $\cos\omega t$ and $\sin\omega t$ are generated correctly ($\cos\omega t$ is cophasal with voltage in phase *a*), the component in axis *d* equals the amplitude of the synchronizing voltage, whereas the component *q* is zero. This property of the abc→dq transformation is employed in the design of the three-phase synchronization circuit depicted in figure 16.
The following designations are used in figure 16: *PI*- proportional-integral controller, *VCO*- voltage controlled square-wave generator. The *PI* controller input signal is the instantaneous value of the *q*-axis component of abc→dq transformation. The controller tunes the *VCO* oscillator, whose output signal controls the $\cos\omega t$ and $\sin\omega t$ generation circuit. The controller

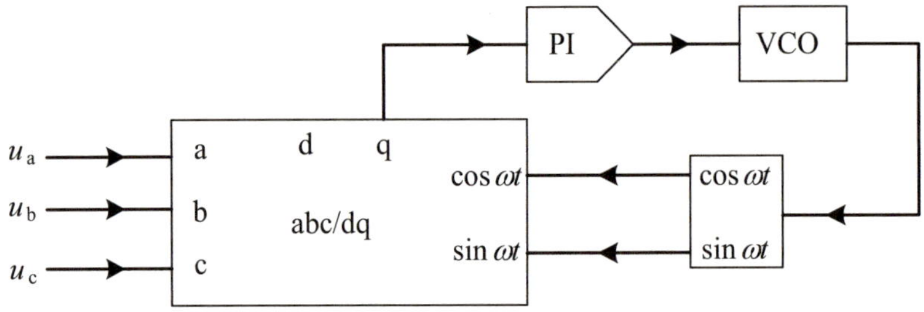

Fig. 16. Block diagram of the synchronization circuit

connected to the q-axis controls the *PI* controller error to zero (the value q-axis component equals zero) what means that, according to equation (4), the generated $\cos\omega t$ signal is cophasal with synchronizing voltage u_a. The control circuit shall attain the state in which the q-axis component value is zero. This condition (q= 0) is satisfied in two cases:

1. The generated function $\cos\omega t$ is cophasal with the synchronizing voltage u_a. This case is described by equation (20). Figure 17 shows oscillograms of the synchronizing voltage (u_a, u_b, u_c) and the generated function $\cos\omega t$. The simulation waveforms are computed using a model of the system implemented in FPGA.

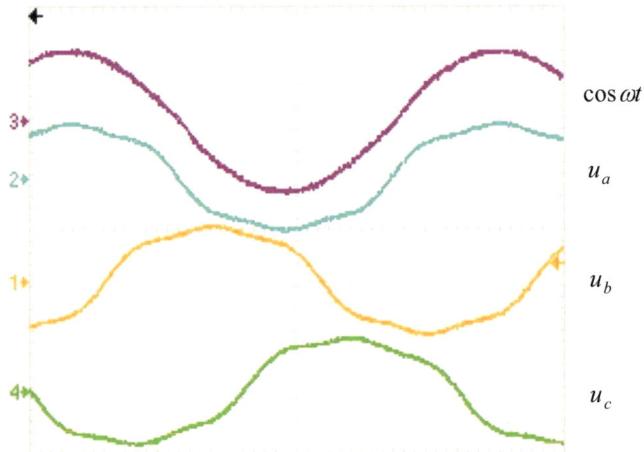

Fig. 17. The synchronizing voltage waveforms and the generated function $\cos\omega t$

2. The generated function $\cos\omega t$ is phase-shifted in with respect to the synchronizing voltage u_a by π. The transformation abc→dq then takes the form described by equation (21).

$$\begin{bmatrix} v_d \\ v_q \end{bmatrix} = \begin{bmatrix} -\cos\omega t & -\sin\omega t \\ \sin\omega t & -\cos\omega t \end{bmatrix} \begin{bmatrix} v_a \\ \frac{1}{\sqrt{3}}(v_b - v_c) \end{bmatrix} \tag{21}$$

Substituting (19) to equation (21) yields (22).

$$\begin{bmatrix} v_d \\ v_q \end{bmatrix} = \begin{bmatrix} -V_m(\cos^2\omega t + \sin^2\omega t) \\ V_m(\sin\omega t\cos\omega t - \sin\omega t\cos\omega t) \end{bmatrix} = \begin{bmatrix} -V_m \\ 0 \end{bmatrix} \tag{22}$$

The phase shift of $\cos\omega t$ function (by π) with respect to the voltage u_a results in erroneous power relationships in the converter operation and is, therefore, inadmissible. This case is illustrated in figure 18 which shows the simulation waveforms computed using the system model.

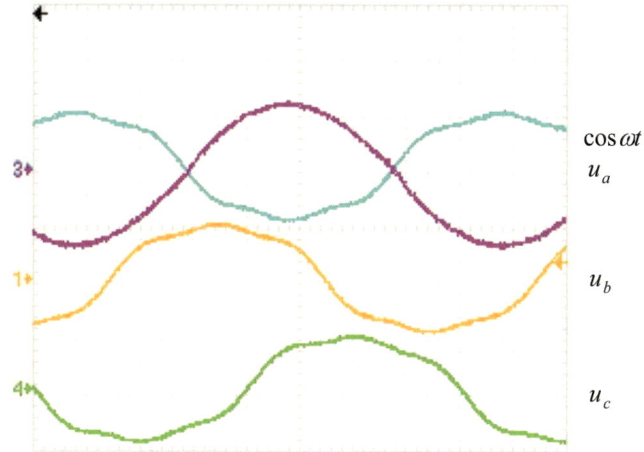

Fig. 18. The synchronizing voltage waveforms and the generated function $\cos\omega t$

The operating point of the synchronizing circuit from figure 16 (case *1* or *2*) is determined by the initial phase of synchronizing voltages at the instant of the converter start. In order to prevent a random selection of the operating point the synchronization circuit should be modified to enforce the system operation according to equation (20) (case*1*). Figure 19 shows two possible versions of the synchronizing circuit.

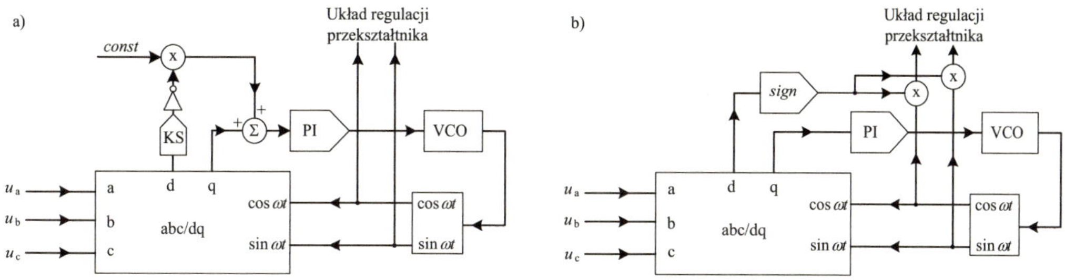

Fig. 19. Two versions of the synchronizing circuit enforcing the correct generation of $\cos\omega t$ function

The following designations are used in figure 19: *const* - a constant value, KS- sign comparator, *sgn*- sign of the signal ("1" if the signal is greater than zero, conversely "-1"), other elements according to figure 14. If the *d*-axis component value (figure 19a) is less than zero, the case described by equation (22), the constant value *const* is added to the instantaneous value in axis *q* (the *PI* controller input error) therefore tuning the *VCO* oscillator, even when the *q*-axis component value is zero. When the control circuit attains the state in which the value of the *d*-axis component is greater than zero only the *q*-axis component is applied to the *PI* controller input (the controller error). In the circuit from figure 6b no modification is made to the synchronization circuit. The functions $\cos\omega t$ and

sinωt are multiplied by the d-axis component sign prior to being applied to the converter control circuit. When the system operates correctly, i.e. according to equation (19), the functions cosωt and sinωt are directly applied to the converter control circuit. When the function cosωt is phase-shifted with respect to the synchronizing voltage u_a by π (equation 22) the multiplication of cosωt and sinωt functions by the sign of d-axis component ("-1") reverses their phases and the correct values of cosωt and sinωt functions are applied to the converter control circuit.
Figure 20 illustrates the synchronization circuit operation, the time graphs are recorded in the laboratory setup using the *SignalTap II Logic Analyzer* tool (a part of the *Quartus II* package).

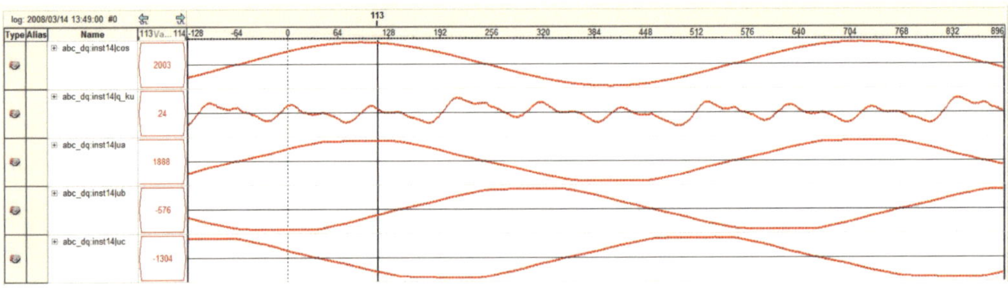

Fig. 20. The time graphs of cosωt function, and the synchronizing voltages waveforms recorded in the experimental setup

Figure 20 shows the oscillograms of phase voltages u_a, u_b, u_c, the generated cosωt function, and q-axis component (the controller in figure 19 input error). In the presented synchronizing circuit implementation all quantities are represented by eleven-bit numbers plus a sign bit. Instantaneous values of these quantities can vary within the range ±2047. According to equation (20) the values of q-axis component are equal zero if the synchronizing voltages are consistent with (19), i.e. are not distorted. Due to the power system voltage harmonic distortion with $1 \pm 6n$ harmonics (where n= 0, 1, 2...) the values recorded in axis q are different from zero. These values were varying over the range ±98, what makes 4.78% of permissible range.

11.2 Investigation of the converter

Figure 21 shows the waveforms of the voltage and current in phase A and the output voltage. As follows from figure 21 the proposed system allows obtaining the rectified voltage value higher than the phase voltage amplitude and forces a sinusoidal current, cophasal with the phase voltage. Figure 22 shows phasor diagrams for several selected parameters of the converter input voltage.
Figure 22a illustrates the case where the source current is cophasal with the supply voltage (cosφ= 1). In figure 22b the source current is lagging (cos$\varphi \neq$ 1), i.e. a reactive component occurs in the source current. Diagrams in figures 22c and 22d are determined for the inverter mode operation (energy is fed back into the power system). As can be seen from the phasor diagrams, the source current fundamental harmonic cosφ (lagging or leading) can be influenced on by means of shaping the converter input voltage (u_{in}) and energy can be fed back into the power system. Example waveforms in figure 23 illustrate the transition from the rectifier to inverter mode of operation (energy is fed back into the power system).

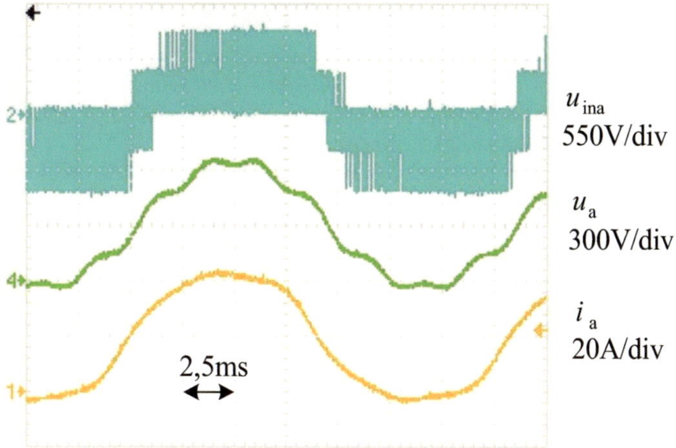

Fig. 21. The waveforms of the phase voltage and current u_a, i_a and the input voltage u_{ina} in phase A

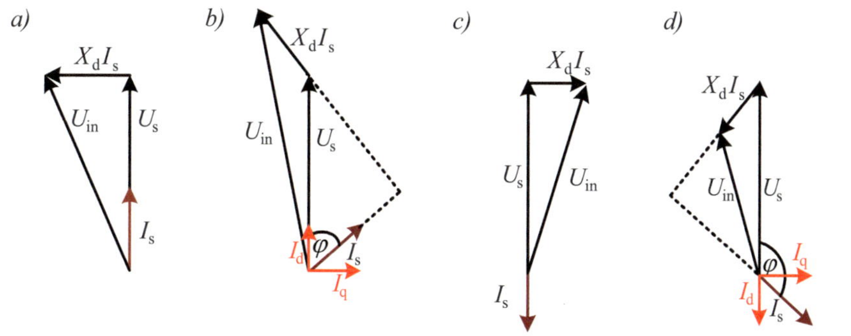

Fig. 22. Phasor diagrams for several selected parameters of the input voltage U_{in} vector

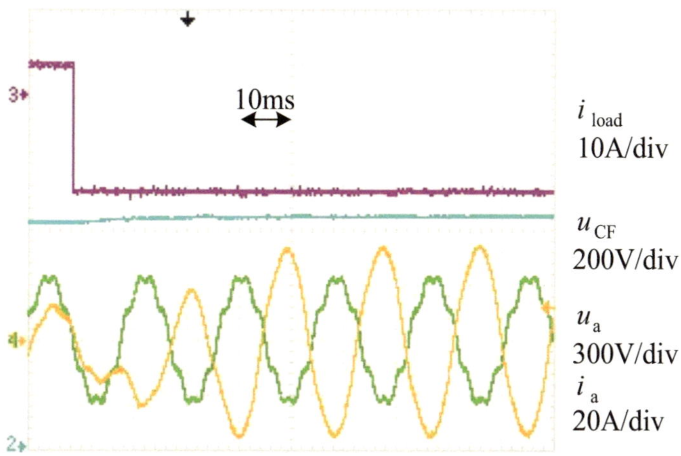

Fig. 23. The waveforms of the load current i_{load}, the output capacitor voltage u_{CF}, the voltage and current in phase A

As follows from figure 23, at the instant the load current direction is reversed, energy is transferred to the capacitor and its voltage u_{CF} increases what, in turn, influences the source current phase shift in such a way that energy stored in the capacitor is fed back into power system.
The described converter is intended for co-operation with an inverter supplying a flywheel energy storage drive. Thus, in order to ensure constant operating conditions of the inverter-motor system, the rectifier should maintain the capacitor voltage at the set value irrespectively of the load current i_{load} value and direction. Figures 14 and 15 show the waveforms recorded at a step change in the load current.
As can be seen from figure 23 at the instant of an increase in the load current the capacitor is discharging (the capacitor voltage decreases) and, consequently, the control system increases the phase current amplitude. Energy is supplied in an amount sufficient to compensate the capacitor voltage decrease resulting from the load current change and provide energy being drawn from the capacitor by the load. When the capacitor voltage becomes equal to the set voltage value of the phase current amplitude decreases to the value which ensures the capacitor voltage is maintained at the required level. In the event of an abrupt change in the load current a reactive component may occur in the phase current, as is shown in figure 25.

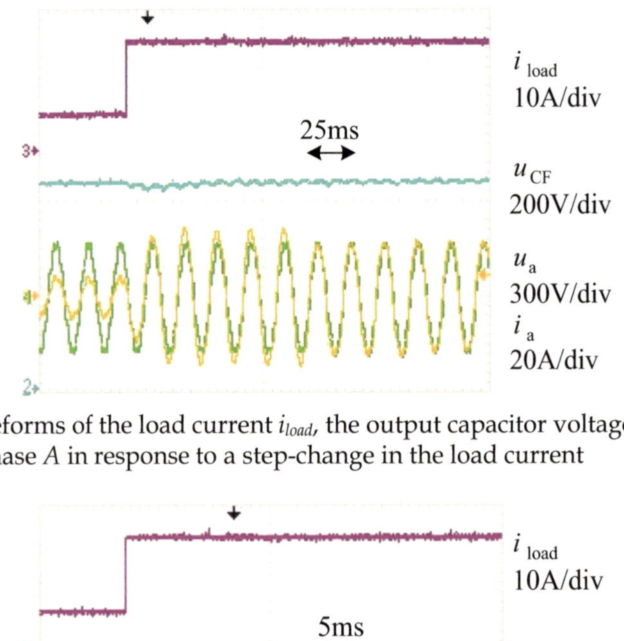

Fig. 24. The waveforms of the load current i_{load}, the output capacitor voltage u_{CF}, the voltage and current in phase A in response to a step-change in the load current

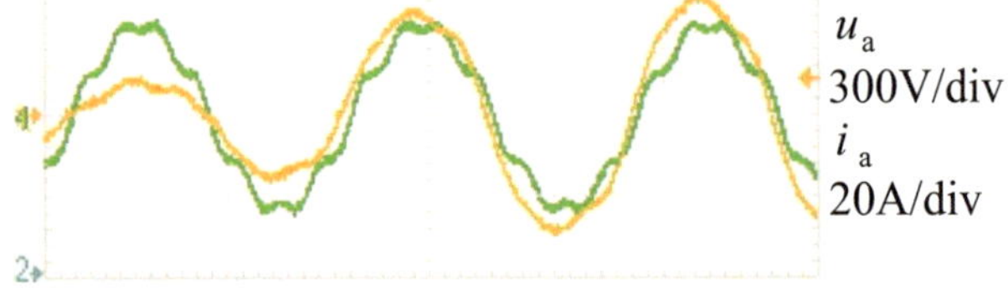

Fig. 25. The waveforms of the load current i_{load}, the output capacitor voltage u_{CF}, the voltage and current in phase A; a reactive component i_q occurs in the source current

12. A flywheel energy storage drive control system

Figure 26 shows the diagram of power processing unit (power supply and inverters) and illustrates the mechanical structure of flywheel energy storage.

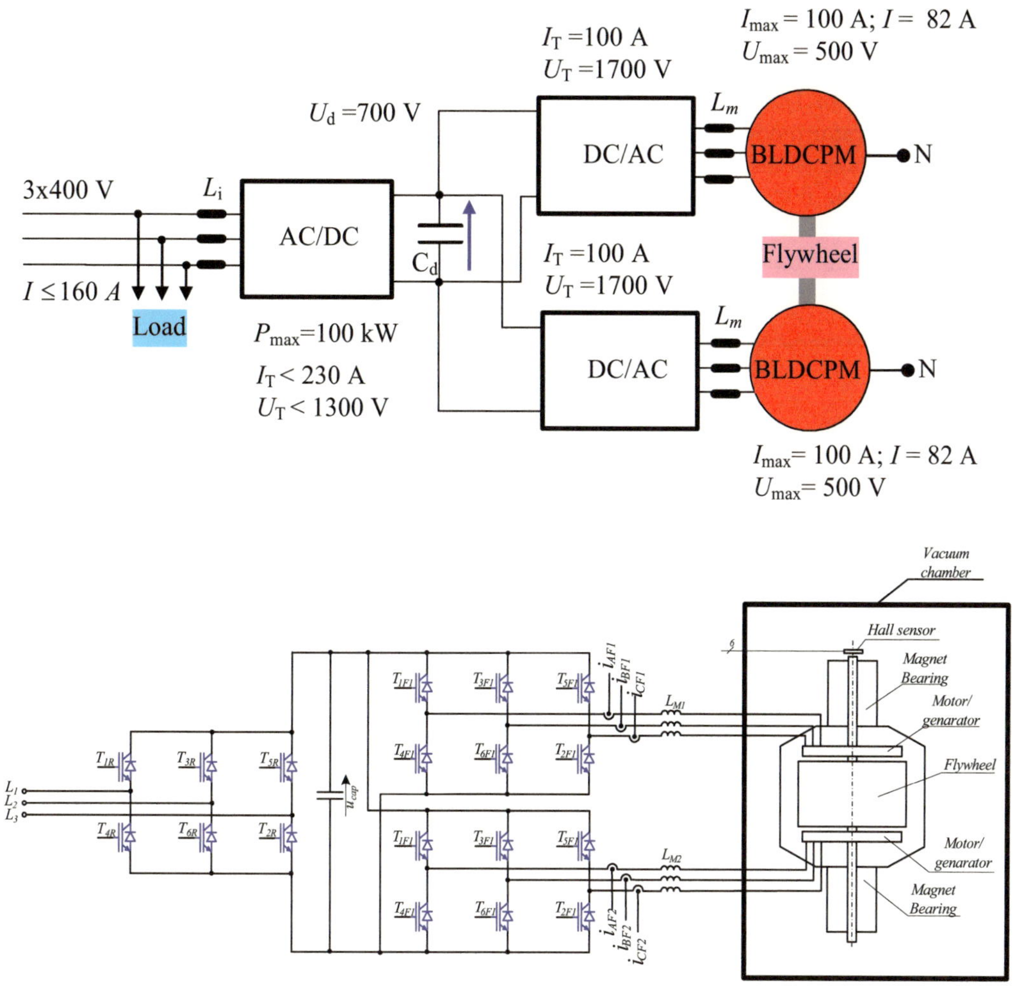

Fig. 26. High-speed flywheel energy storage: a - block diagram, b - power processing unit diagram

Two brushless permanent magnet DC motor are mounted on a common shaft. Magnetic bearings levitate the spinning mass in order to minimize the resistance to motion. The whole structure is enclosed in a vacuum chamber. The motor shaft positioning with respect to the stator is achieved by means of Hall sensors. they determine the instances of switching the inverter switches based on the actual rotor position relative to stator windings axes. The layout of magnets and windings of both motors is the same, thus the instants of commutation are determined by means of a single sensor, common for both motors. Each of the two FES motors is supplied from an independent inverter, of identical structure. Figure 28 shows diagram of the inverters control system.

The following symbols are used in figure 28: H_A, H_B, H_C – the Hall encoder signals (the rotor position with respect to stator), *MUX* - multiplexer, *SAW* - symmetric sawtooth signal generator, u_{cap} – DC link capacitor voltage, *ABS* – absolute value, *KS* – sign comparator, i_{xFy} – (x= *A, B, C;* y= *1, 2*) the motors phase currents. The PWM generators sawtooth signals of both motors are shifted by *T/2*; consequently, the DC link capacitor current alternating component is doubled thereby reducing torque ripples at the FES shaft.

The converter controls the DC link capacitor voltage and if it drops below the predefined level the speed of 1/3 ω_{max} is set in the control system. The system turns from the motor mode to generator mode and the mechanical energy is converted into electrical energy. The capacitor voltage is also controlled during the converter start-up. The inverters remain blocked until the instant of a correct start-up of the line-side converter. The motor actual rotational speed is determined from the frequency of the Hall sensor signals. An algorithm for computing the frequency is described further below.

Fig. 27. Photograph of the FES mechanical structure

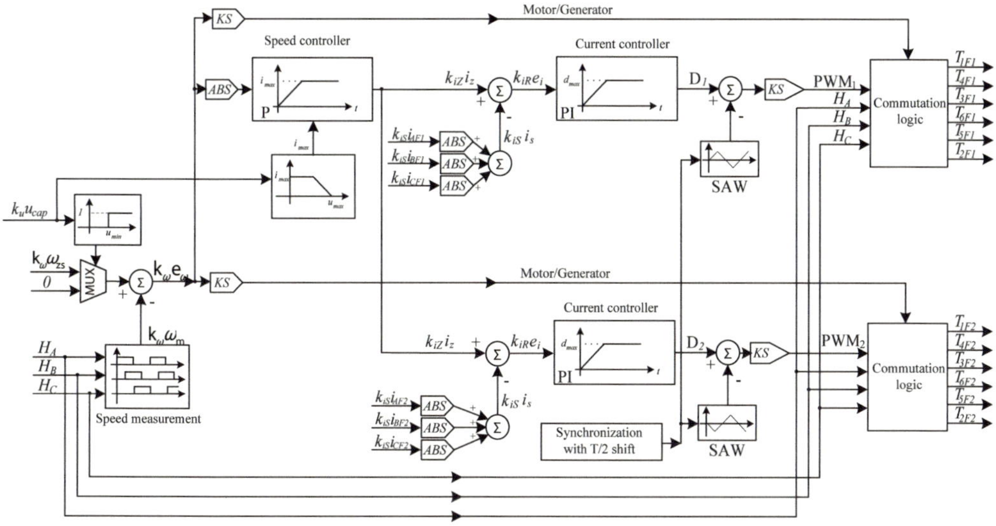

Fig. 28. The inverters' control system

The motors operate at a common shaft and rotate with same speed, therefore the inverters control system (Fig. 28) utilizes a single speed controller, common for both inverters. The speed controller output signal is proportional to the drive current reference value. In order to protect the system against an uncontrolled increase in the DC-link capacitor current the controller output signal is limited to a selected value proportional to the capacitor maximum voltage. This limitation determines the inverters' maximum current. When motors are operated in the generator mode the current limit level must be reduced depending on the instantaneous value of the DC-link capacitor voltage. When the energy recovered from the spinning mass (i.e. delivered to the capacitor) is larger than that supplied by the line-side converter to the supply line, the capacitor voltage increases. An increase in the capacitor voltage results in reduction of the speed controller output limit (i_{max}) and thereby the drive current limiting. If the capacitor voltage reaches its maximum permissible value the set current decreases to zero. The principle is illustrated in figure 29.

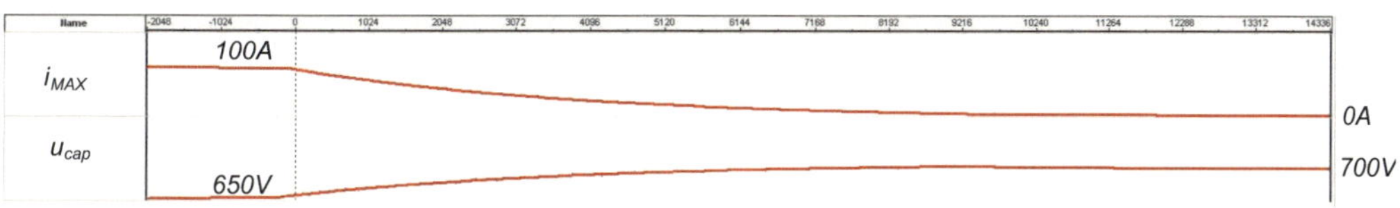

Fig. 29. The DC-link capacitor voltage (alternate component) and the speed controller output limitation (maximum permissible inverter current) versus time

Direction of energy transfer: power supply network ⇒ spinning energy storage (the inertial element ⇒ power supply network) is determined by the sign of the speed control error. If the set speed is lower than the actual rotation speed, the motors turn into generating mode (regenerative braking). In the alternate case the control error is grater than zero, the motors are accelerated or energy supplied from the power network compensates losses resulting from the resistance to motion. The control system automatically sets zero speed if the DC-link capacitor voltage is lower than 0.9 of its nominal value. This condition limits the supply line current during the line-side converter start-up. The drive is started only if the capacitor is charged to $0.9U_N$ with delay of 2 seconds. In the event of voltage loss in the supply line, resulting in the capacitor voltage reduction, the system automatically turns to the generator mode.

The speed controller output signal (the current reference) is compared with the sum of absolute values of the motors phase currents. The inverters' current error ($k_{iR}e_i$) is applied to the PI current controller. Each inverter is provided with an independent controller divided into two parallel components: the proportional and integral part. Both parts of the controller have their own limits. The controller integrator incorporates a limiter that prevents counting when the integer value reaches a predefined maximum level. A separate limitation at the controller output prevents reaching the output signal values that cannot be executed by the control circuit. This limitation results from PWM generators operation area range.

The block termed "Commutation Logic", shown on the diagram in figure 3 is responsible for correct switching of the inverters' transistors, depending on the permanent magnets position with respect to the stator windings. Time relations between the motor electromotive force (e_A, e_B, e_C), Hall sensors signals (H_A, H_B, H_C), transistor switches control pulses (T_1, $T_2 \div T_6$) and phase currents (I_A, I_B, I_C) are shown in figure 30 (Fig. 30a refers to the motor mode operation, Fig. 30b refers to the generator mode). As can be seen from figure 30, logic functions controlling the switches in the motor and generator mode operation are different (transistor gate control pulses are shifted by $T/2$). The "Commutation Logic" block structure is shown in Fig. 31 The "Pulse Blocking" input is employed for blocking all transistors during starting (until the capacitor voltage reaches $0.9U_N$) and to turn off the line-side converter upon detection of exceeding the current permissible value.

The position sensor shall change its logic state at the angular distance of $\pi/6$ from the motor phase voltage (e_A, e_B, e_C) zero crossing. The current flows always through windings in which maximum voltage value occurs (Fig. 30). Therefore, during a full revolution of the rotor each of the inverter's transistors conducts during $2\pi/3$ of the cycle and participates in two from the six allowable pairs: T_1T_2, T_2T_3, T_3T_4, T_4T_5, T_5T_6, T_6T_1. Since only two switches can conduct simultaneously, in order to minimize switching losses only one transistor of a pair is chopper controlled while the other is continuously turned on. To ensure uniform heating of a transistor module a transistor is continuously turned on during ½ of the conducting period, while during the other half it is chopper controlled. In figure 30 the transistor switching process is indicated by shaded area. Figure 31a depicts the rotor position sensor signals and transistors gate control pulses (the waveforms recorded in Quartus II programme using the *SignalTap II Logic Analyzer* tool). Figure 31b shows oscillogram of the phase current waveform and a transistor gate control pulses. From the principle of inverter operation it follows that conduction times (pulse duty factor) of all the inverter transistors are the same. The PWM generator module operates in a continuous manner; the commutation logic, shown in figure mb.6, is responsible for assignment of control signals to individual transistors (depending on the rotor position with respect to stator).

In this figure each transistors has assigned its individual control signal PWM_x ($x = 1, 2 \ldots 6$), which is a logical function of the common control PWM and the position sensors pulses; the functions for both the motor and generator operation mode are listed in table 2. The separation of the common control and the use of an appropriate logical function allows limitation of transistors switching losses according to the idea illustrated in figures 30 and 31. Logical functions listed in table mb.1 and logical circuits from figure mb.6, are exclusively correct for the phase sequence shown in figure 30.

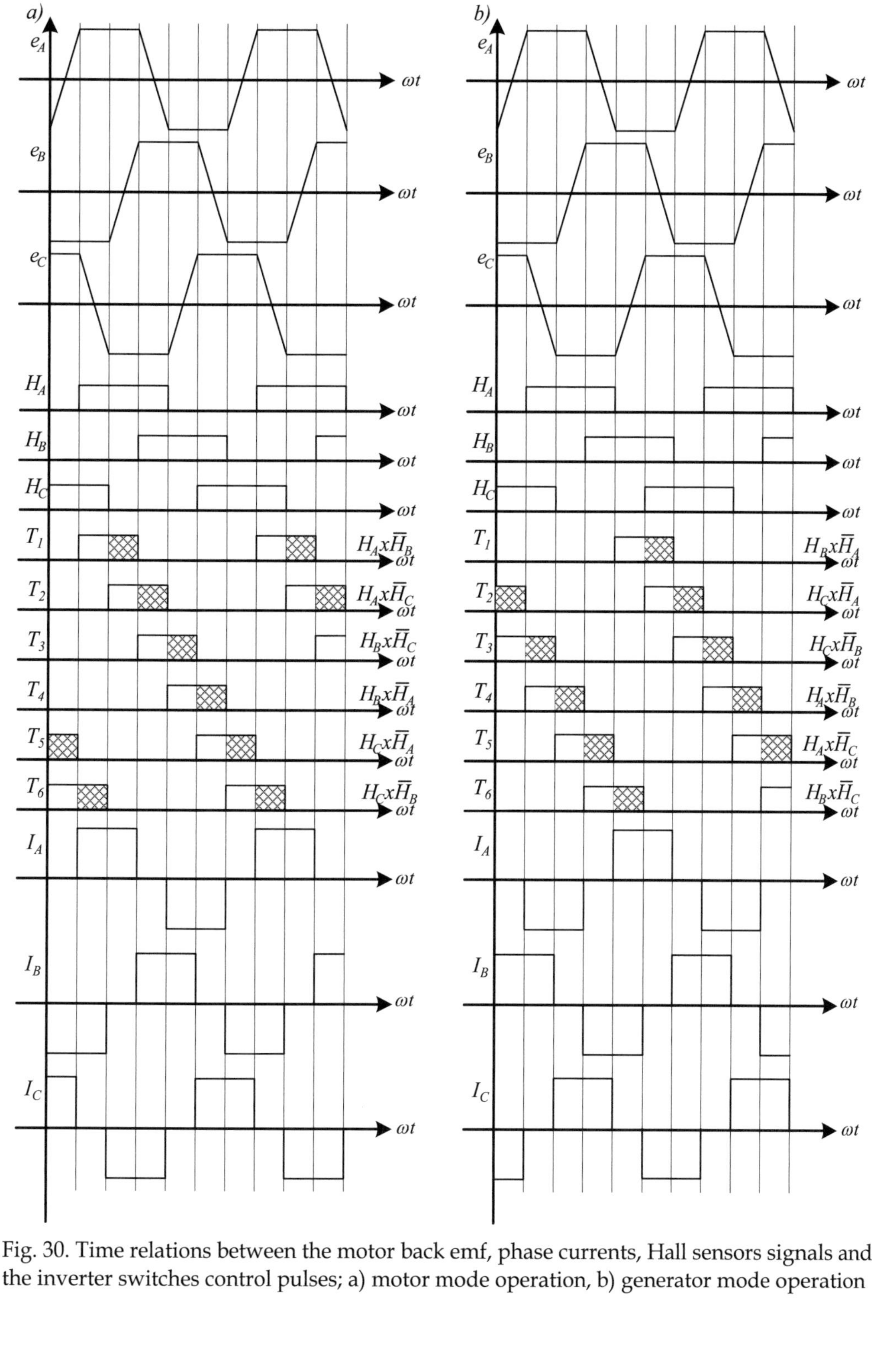

Fig. 30. Time relations between the motor back emf, phase currents, Hall sensors signals and the inverter switches control pulses; a) motor mode operation, b) generator mode operation

Transistor	Motor	Generator
T_1= PWM_1	PWM **or** (H_A **and** H_C)	PWM **or** (**not**(H_A) **and not**(H_C))
T_2= PWM_2	PWM **or** (**not**(H_B) **and not**(H_C))	PWM **or** (H_B **and** H_C)
T_3= PWM_3	PWM **or** (H_A **and** H_B)	PWM **or** (**not**(H_A) **and not**(H_B))
T_4= PWM_4	PWM **or** (**not**(H_A) **and not**(H_C))	PWM **or** (H_A **and** H_C)
T_5= PWM_5	PWM **or** (H_B **and** H_C)	PWM **or** (**not**(H_B) **and not**(H_C))
T_6= PWM_6	PWM **or** (**not**(H_A) **and not**(H_B))	PWM **or** (H_A **and** H_B)

Table 2. Logical functions for individual transistors control that allow minimizing switching losses

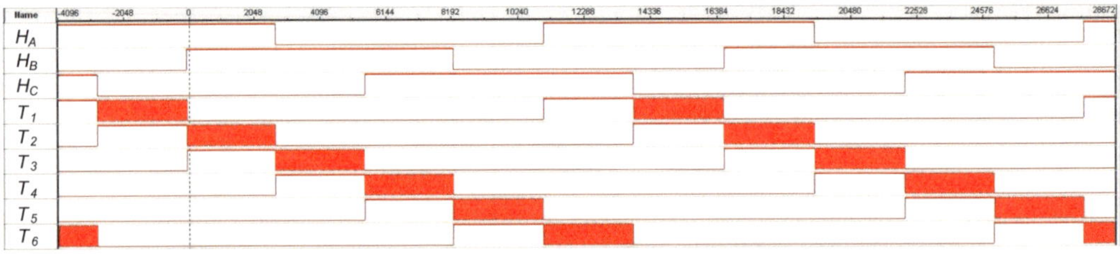

(a)

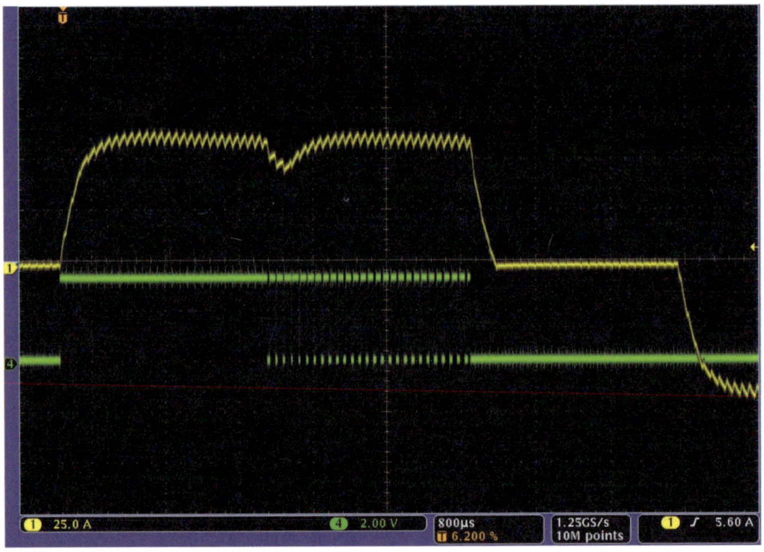

(b)

Fig. 31. a) Hall sensors signals and transistors gate control pulses; b) The motor phase current and transistor gate control pulses limiting switching losses

13. Speed measurement

The converter control system utilizes signals from rotor position sensors to detect which winding conducts current and, basing on their frequency, determines the speed of FES rotation. Using these signals the control algorithm can detect sensor failure (loss of connection integrity) or locked rotor (because of e.g. bearings failure). If the logical state in all three signal lines (*HA, HB, HC*) does not change over a specified time interval, transistors' control pulses are blocked. This action protects the motor windings against overheating due to continuous current conduction. This blocking is independent of the speed measurement because the algorithm of digital speed measurement assumes a minimum determinable speed value, whereas the described failure detection method works correctly also at arbitrary low speeds. This is of particular importance when motors are started from zero speed.

The rotational speed measurement algorithm shall ensure the frequency of the output signal to be as high as possible and possible misalignment shall not impact the measurement result. Ideally, in a theoretical case, the position sensor pulses (*HA, HB, HC*) duty factor is 50%. If the Hall sensor axis is misaligned with respect to the motor shaft axis, or the sensor is not mounted perpendicularly to the shaft, the duty factor differs from the required value. The cycle of each of the three rotor position signals equals one period of the shaft revolution (for a two-pole pair motor) or a half of the revolution period - for a four-pole motor. Since the position sensor signals are shifted with respect to each other by 1/3 T, then determining the cycle of each signal (using either a rising or a falling edge) the measurement frequency is three times the rotation frequency (for a two-pole motor). The speed measurement frequency can be doubled determining the signal half-cycle (using both the rising and falling edge) and employing a supplementary register that stores the determined value of the preceding half-cycle. For each change in the sensor signal level the determined signal period is the sum of the current measurement result and that being stored in the supplementary register. The flowchart of the algorithm determining the sensor signal period is shown in figure 32.

The algorithm from figure mb.9 is executed in an infinite loop independently for each of three signals. Each step is executed on the rising edge of the clock signal (CLK) of known frequency. The internal counter representing the revolution period is incremented by one at the clock rising edge (the variable *Counter* in Fig. 32), next the level of the Hall sensor signal (*HA, HB, HC*) is checked. A change of logical state is interpreted as completing 1/2 of shaft revolution (1/4 for a motor with two pole-pairs). In such a case in supplementary registers (*R2_x, R1_x* where *x*= *A, B,* C) the current and the preceding value of counter is stored. The sum of the *R2_x, R1_x* registers values represents the rotational speed (frequency computed from completing a full revolution). If the sensor logical state did not change the counter value is checked and when it attains the specified maximum value it is assumed that the motor is stopped and its rotational speed is zero. In the case of a motor with two pole-pairs four supplementary registers are required, each of them stores the duration of the consecutive fourth part of a revolution. Figure 33 shows the waveforms recorded in Quartus II programme illustrating practical realization of the described algorithm. The variable *Counter* (Fig. 32) has not been taken into account in the practical realization, is function is fulfilled by the register *R1_x*.

The following symbols are used in figure 33:

H_A, H_B, H_C - pulses from the rotor position sensor determining the current commutation instants (e.g. Hall sensor, sensorless method),

G_A, G_B, G_C – signals of the rotor position sensors state change (pulse edge detection),

G_I – the speed controller timing signal.

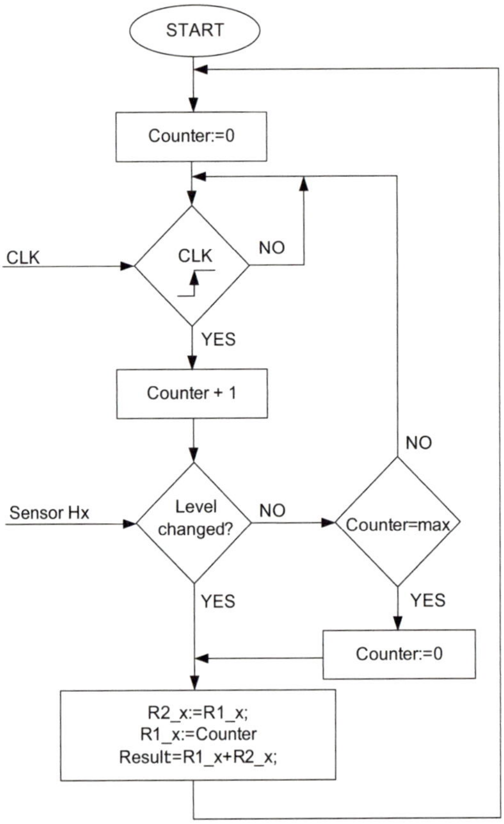

Fig. 32. Flowchart of the algorithm determining the frequency of a single Hall sensor signal

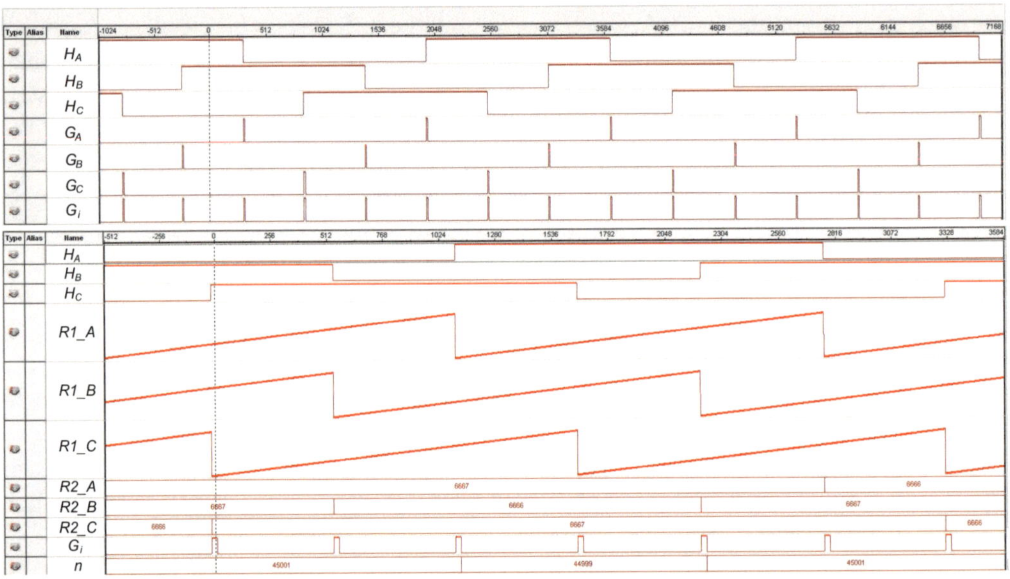

Fig. 33. The principle of the rotational speed measurement – a practical realization

Duration of the position sensor high and low states (a half of a revolution for a motor with one pair of poles or a quarter of a revolution for a motor with two pole-pairs) is timed by counting pulses of a known frequency between the consecutive pulses *Gx* (x= *A, B, C*). The rising edge *Gx* resets the *R1_x* (x= *A, B, C*) register counter (Fig. 27). Prior to resetting the *R1_x* register, its value is rewritten to the supplementary register *R2_x*. Knowing the reference signal frequency (f_{CLK}) and states of the registers *R1_x* and *R2_x* at falling edge of the *GI* signal we can determine the motor rotational speed from equation (23)

$$n = \frac{60 f_{CLK}}{R1\_x + R2\_x} \tag{23}$$

14. References

Aanstoos, T.A. Kajs, Brinkman J. P., Liu W.G., Ouroua H. P., Hayes A., Hearn R.J., Sarjeant C., Gill J., H.: *High voltage stator for a flywheel energy storage system*; Center for Electromech., Texas Univ., Austin, TX; Magnetics, IEEE Transactions on 04/25/2000 -04/28/2000, Jan 2001 Location: San Francisco, CA, USA

Acarnley P.P., Mecrow B. C., Burdess J. C., Fawcett J. N., Kelly J. G., Dickinson P. G.: *Design Principles for a Flywheel Energy Store for Road Vehicles*, IEEE Transactions on Industry Application vol. 32, no 6, 1996,

Akagi H., Sato H.: *Control and performance of a doubly-fed induction machine intended for a flywheel energy storage system*, IEEE Transactions on Power Electronics, Volume: 17 Issue: 1, Jan 2002,

Curtiss, D.H. Mongeau, P.P. Puterbaugh, R.L.: *Advanced composite flywheel structural design for a pulsed disk alternator*; Kaman Electromagnetics Corp., Hudson, MA; Magnetics, IEEE Transactions on 04/20/1994 -04/24/1994, 20-24 Apr 1994 San Diego, CA, USA.

Davies T. A., Jefferson C. M.: *Wind-power flywheel integration* Dept. of Eng., Bristol Polytech: Energy Conversion Engineering Conference, 1989. IECEC-89. Proceedings of the 24th Intersociety 08/06/1989 -08/11/1989, 6-11 Aug 1989 Washington, DC, USA; page(s): 2071-2076 vol. 4.

Davies, T.S. Larsen, N. *A: Regenerative drive for incorporating flywheel energy storage into wind generation systems*; Energy Conversion Engineering Conference, 1989. IECEC-89. Proceedings of the 24th Intersociety 08/06/1989 -08/11/1989, 6-11 Aug 1989 Washington, DC, USA; page(s): 2065-2069 vol.4 6-11 Aug 1989.

Ginter, S. Gisler, G. Hanks, J. Havenhill, D. Robinson, W. Spina, L. SatCon Technol. Corp., Tucson, AZ; *Spacecraft energy storage systems:* IEEE Aerospace and Electronics Systems Magazine; page(s): 27-32 Volume: 13, May 1998.

Gosiewski Z., Kulesza Z.: *Wykorzystanie układów FPGA do sterowania siłowników łożyska magnetycznego.*, "Pomiary, Automatyka, Robotyka" R.11, nr 2/2007.

Gully, J.H. Pratap, S.B. Headifen, R.N. Marinos, C. Dick, W. Goodel, B.: *Investigation of an alternator charged pulse forming network with flywheel energy storage* Center for Electromech., Texas Univ., Austin, TX; Magnetics, IEEE Transactions on 04/28/1992 -04/30/1992, 28-30 Apr 1992.

Hall C. D.: *High Speed Flywheels for Integrated Energy Storage and Attitude Control*. Department of Aeronautics and Astronautics Air Force Institute of Technology/ENY Wright-Petterson Air Force Base, Ohio 45433, Sept. 24, 1997.

Hebner R., Beno J., Walls A.: *Flywheel Batteries Come Around Again*, IEEE Spectrum April, 2002.

Kamiński G., Szczypior J., Biernat A., Smak A., Rowiński A., Kowalski W.: *Konstrukcja modelu maszyny do elektromechanicznego magazynowania energii*, Przegląd Elektrotechniczny 6, 2008.

Kamiński G., Szczypior J., Smak A.: *Analiza kształtu zwojów uzwojeń twornika bezrdzeniowego w maszynach do kinetycznych magazynów energii*. Przegląd Elektrotechniczny LXXXI, 10, 2005, str. 7-14.

Kamiński G., Szczypior J., Smak A.: *Analiza uzwojeń tworników bez rdzeni ferromagnetycznych w maszynach kinetycznych magazynów energii*, SME Opole 2005.

Kamiński G., Szczypior J., Smak A.: *Wybrane problemy konstrukcji maszyny bezszczotkowej stosowanej w kinetycznym magazynie energii*. SME Opole 2005.

Kulesza Z., Gosiewski Z.: *An FPGA implementation of the robust controller for the active magnetic bearing system*, 4th International Conference "Mechatronic Systems and Materials, MSM 2008". Abstract book. Bialystok 2008. s. 126.

Kulesza Z., Gosiewski Z.: *An FPGA implementation of the robust controller for the active magnetic bearing system*. "Solid State Phenomena", Vol. 147-149 (2009), s. 399-409.

Mason L.: *Flywheel Power System*, NASA Lewis Research Center, Flywheel Workshop, Oct. 7, 1998.

Mulcahy, T.M. Hull, J.R. Uherka, K.L. Abboud, R.G. Juna, J.J.: *Test results of 2-kWh flywheel using passive PM and HTS bearings*, Argonne Nat. Lab., IL; Applied Superconductivity, IEEE Transactions on 09/17/2000 -09/22/2000, Mar 2001.

Piróg S., Siostrzonek T., Penczek A., Czekoński J.: *Wirujący zasobnik energii z bezszczotkowym silnikiem prądu stałego - wyniki badań laboratoryjnych* VIII Konferencja Naukowa „Sterowanie w Energoelektronice i Napędzie Elektrycznym" SENE'2007 (pod auspicjami European Power Electronics and Drives Association) Pol. Łódzka Łódź 21-23 listopad 2007 (materiały konferencji t. II, str. 379-385..

Piróg S., Siostrzonek T., Penczek A.: *The flywheel energy storage with brushless DC motor* International conference on Power ELectronics and Intelligent Control for Energy Conservation: Warsaw, Poland, October 16-19, 2005.

Piróg S.: *Elektromechaniczne magazyny energii* Napędy i Sterowanie nr. 12 2006, str. 115-132.

Piróg S.: *Magazyny energii* SENE 2007: VIII krajowa konferencja naukowa Sterowanie w Energoelektronice i Napędzie Elektrycznym, Łódź, 21-23 listopada 2007 r.

4

Energy Storage in Grid-Connected Photovoltaic Plants

Rosario Carbone
University "Mediterranea" of Reggio Calabria
Italy

1. Introduction

Nowadays, photovoltaic (PV) plants are receiving a very great attention due to their intrinsic ability to directly transform solar energy in electrical energy.

Nevertheless, electricity generated from photovoltaic plants can rarely provide immediate response to load demand, as these sources do not deliver a regular supply immediately compatible with consumption needs.

In stand-alone PV plants, energy storage (typically based on electrochemical batteries), together with the help of additional generation systems (such as those powered by fuel engines), is on the basis of regularization of PV generation and of full satisfaction of load consumptions.

In grid-connected PV plants – theoretically - energy storage is not necessary or useful, due to the availability of the distribution grid that should work as an ideal container of the electrical energy (theoretically, it can work both as an ideal generator and, also, as an ideal load). However, in this last years, an important attention has been devoted to the use of energy storage also in grid-connected PV plants, with the main aim of overcoming some important power quality problems of real distribution grids and for making PV plants more and more useful and attractive.

In fact, avoiding more extensive details, energy-storage seems able to improve reliability, availability and energy generation efficiency of grid-connected PV plants, still poor because a lot of problems including: the high variability of solar energy availability, the high energy generation losses (i.e. caused by non-uniform solar irradiation conditions of PV fields) and the disturbances introduced by non-linear behaviour of power electronic apparatus (widely, PWM inverters) currently utilized for interconnecting PV plants with distribution grids.

Most of these recent studies are substantially focused on sizing, operating and analysing of electrochemical energy storage systems to be used as an "energy buffer" for reducing, within acceptable limits, aforementioned problems and for improving some important grid auxiliary services as:

- load balancing and peak shaving;
- compensation of disturbances (outages, voltage drops and dips, PF correction, …);
- economic optimization of generated/consumed energy flows;
- optimal integration among different kind of renewable energy resources (PV, wind, fuel-cells, …).

Aforementioned ancillary services are undoubtedly of strategic importance for future large-scale development and diffusion of grid-connected PV plants.
However, as evidenced in current researches, they are based on centralized and conspicuous electrochemical amount of batteries (based on different technologies) whose potentials, in the opinion of the author, can be further improved [Denholm et al., 2007; Lo et al., 2005; Lu and Shahidehpour, 2005].
Having in mind the objective to obtain, in addition to previously mentioned ancillary services, the important goal of improving reliability, availability and electrical energy generation capability of grid-connected PV plants (under assigned solar irradiation conditions), the new idea that will be fully analysed in this work is that batteries can be used in a new and more effective "distributed manner".
In this new context, batteries, with a reduced whole capacity value, can be utilized as distributed passive maximum power point trackers (MPPT) able of maximizing the amount of energy generated by grid-connected PV plants in presence of non-uniform shadowing (partial shadowing) effects on PV fields, as an alternative to the more expensive and complex active MPPTs [Carbone, 2009].
In the following, after introducing the potential benefits of the electricity storage in distribution grids (section 2) and after recalling currently utilized methodologies for improving power generation of grid-connected PV plants (section 3), the proposal of utilizing the energy storage in grid-connected PV plants, operated by using batteries in a distributed manner, is fully described (section 4) from the theoretical point of view.
In section 5, the effectiveness and the usefulness of the propped methodology is fully investigated, by means of a lot of experimental tests.
In section 5.1, experimental tests are performed and discussed by referring to a small-scale prototype of about 20 Wp of power, constituted by 4 PV small-panels, each of about 5 Wp, together with 8 NiCd rechargeable batteries, each with 1.2 V and 800 mAh of capacity.
In section 5.2, experimental tests are performed and discussed by referring to a physically realized 18 kWp PV plant, connected to the LV distribution grid and owned by the B&T Solar Energy Division Company, on Lazzaro (RC), Italy. The PV plant is realized by means of 90 PV modules with 200 Wp and with an open circuit voltage of about 40 V; PV modules are grouped in 3 PV sub-fields (with 30 modules and 6 kWp) connected to the grid by means of three different single-phase PWM inverters. Because of each inverter has two separate DC-inputs, with separate DC-DC boost-converter operating the MPPT function, each sub-field is realized with 2 different groups of 15 series-connected PV modules. For our experimental tests and measurements, only one group of 15 series-connected PV modules (3 kWp) of a single-phase PV sub-field has been considered; furthermore, because of the temporarily limited availability of batteries, only 9 modules of the 15 series-connected PV modules have been experimented with (and without) batteries. In particular, a group of 3 series-connected lead-acid batteries, each with 12V and 10 Ah of capacity, are connected in parallel to each group of 3 parallel-connected PV modules.
Finally, in section 6, some perspectives on the possibility of designing and developing single AC PV modules, with on board a distributed energy storage system, for direct connection with low voltage distribution grids and characterized by high availability and high quality of the output AC voltage waveform, are also developed.

2. Advantages of energy storage in electricity distribution grids

In recent years, a lot of papers have been published in the specialised literature to discuss about the usefulness of energy storage systems in modern distribution systems, especially in presence of distributed generation systems from renewable (PV plants, wind farms, fuel cells, high-efficiency co-generation systems, ...).

Regardless of their type and their technology (batteries, super-capacitors, super-magnetics, flywheels, pumped-hydro, compressed air, hydrogen, ...), the main outcome of all these studies is that energy storage systems are on the basis of the achievement of very important and valuable ancillary services, able to significantly improve reliability, availability and power quality of modern distribution grids.

A first basic concept is that the energy storage adds value to distributed generation plants from renewable by making them predictable.

That said, one of the most critical problems of distribution grids is, undoubtedly, that of satisfying the peaks of the load demand. The use of one ore more of the aforementioned energy storage systems can effectively compensate for load variations, so making possible to operate transmission, sub-transmission, and distribution networks with lighter designs, that is to say, energy storage can be used instead of more complex, expensive and inefficient needed solutions (over-sizing of base-load generation units, peaking generation units based on combustion turbines, ...).

Energy storage facilitates the active and reactive power flow control for distribution grid voltage regulation.

Energy storage at power plants may provide "black-start" capability (power for plants that need electricity to start up).

Energy storage may have special use in applications such as momentary carry-over for short outages to high-value industrial processes and/or plants, voltage support, power factor correction and other aspects of power quality.

Energy storage facilitates the interconnection among different generation plants from renewable (solar, wind, fuell-cells, ...) and may make them more reliable and efficient.

In a power system where distribution grids provide an energy storage, base-load generation units (typically, coal or nuclear power plants) could properly and efficiently operate at the full power for all time, with the surplus of generated energy being available for charging the energy storage system.

Taking advantage of the new contest of the free market of the electrical energy, in a distribution grid with energy storage systems, electrical energy can be purchased during a low load demand, at low rates, can be stored and, then, can be sold, during a peak of the load demand, at a higher rate.

The American Electric Power Company (AEP) ranks among the nation's largest generators of electricity, owning nearly 38.000 megawatts of generating capacity in the U.S. In the recent years, AEP studied the direct and indirect benefits, strengths and weaknesses of distributed energy storage systems (DESS) and chose to transform its entire utility grid into a system that achieves optimal integration of both central and distributed energy assets. To that end, AEP installed the first NAS battery-based energy storage system in North America. After one year of operation and testing, AEP has concluded that, although the initial costs of DESS are greater than conventional power solutions, the net benefits justify the AEP decision to create a grid of DESS with intelligent monitoring, communications, and control, in order to enable the utility grid of the future [Nourai & Kearns, 2010].

3. Losses and power generation improvements in grid-connected PV plants

3.1 Losses

In this section the most common losses on grid-connected PV plants will be briefly recalled, in order to better understand current and future solutions that can be used to mitigate this important problem.

3.1.1 Losses caused by the mismatching between the PV plant and the grid

One of the most critical problem for grid-connected PV plant is, undoubtedly, that of the mismatch, that can occur at a given time, between the power injection capability of the PV plant and the power absorption capability of the grid. By considering the grid fully passive and resistive, for the sake of simplicity, this problem can be appreciated with the help of Fig.1.(a), in which the working point of the system composed by the PV plant and the grid is graphically found by superimposing the I-V characteristic of the PV plant with the I-V characteristic of the grid. In practice, at a given time, the said working point can be significantly different from that in which the power exchanged between the PV plant and the grid is equal to the maximum power that the PV plant could generate (MPP). Furthermore, the working point changes if the equivalent resistance of the grid changes and/or if the solar irradiation (SI) of the PV field changes. Obviously, this phenomenon - if not controlled - may cause significant generating power losses.

3.1.2 Losses caused by mismatching conditions among PV modules of the PV plant

Important mismatching in working conditions of different PV modules of the same PV plant can occur, essentially because of:

- discrepancies in module parameter values caused by manufacturing tolerances,
- different module ageing effects,
- different orientations of modules.

Furthermore, especially in a urban residence, where PV modules of a PV plant are normally installed on the roofs, in addition to shadows created by clouds, those created by neighbouring buildings, trees, power and/or telephone lines, sometimes, partially cover PV modules (partial shadowing), so creating a condition of mismatching or "unbalanced generation" among different PV modules.

It is, also, well known that, in order to allow the utility-interactive power electronic inverter to obtain sufficient DC voltage, a lot of PV modules are usually connected in series, so making up a so called "PV string".

In case of mismatching conditions, the current generated by a PV string is physically imposed to be equal to the minimum current value, corresponding to the current generated by the most unfavourably irradiated PV module of the PV string, and, finally, the whole power generated by the PV string can be unfavourably reduced with respect to the maximum available power. In other words, a significant reduction in the total output power of the PV system is often observed under the aforementioned mismatching conditions. In order to alleviate the aforementioned problem, traditionally, a bypass diode is connected in parallel with each PV module; this alleviates the power reduction of the PV string, however, the shaded PV modules cannot generate their inherent power, and hence the problem has no its optimal solution.

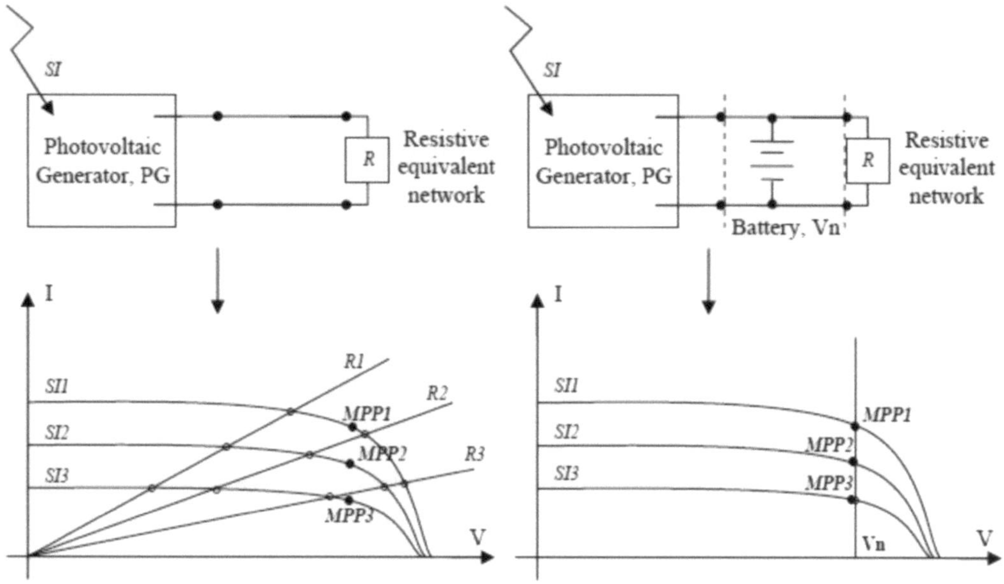

(a) Simplified scheme and I-V characteristics of a conventional grid-connected PG

(b) Simplified scheme and I-V characteristics of grid-connected PG in presence of batteries

Fig. 1. Working points analysis of grid-connected photovoltaic systems, without (a) and with (b) the use of batteries.

3.1.3 Losses caused by the reversed power flow on grids

Radial power distribution grids are designed for a power flow from the high voltage (HV) side to the low voltage (LV) side.

In presence of Grid-connected PV plants, power flow is reversed from the end of LV side to the HV side of grids and his may cause voltage rise of LV line. To prevent the over voltage of the power distribution lines, power electronic inverters for PV plants (i.e. for Japanese PV plants [Ueda et al., 2006]) have a function that regulate the output power when the voltage of the grid is too high; because of this function, significant amount of electric power of PV plants can be lost [Ueda et al., 2006].

3.2 MPPT techniques

After discussing about power losses on grid-connected PV plants, some of the most diffused techniques, utilized for mitigating them, are now recalled.

In general terms, techniques for optimising power generated by PV plants are known as maximum power point tracking (MPPT) techniques.

3.2.1 Active MPPTs

As well known, losses evidenced in section 3.1.1 are usually and effectively avoided by implementing on power electronic interfacing apparatus (inverters) control logics able to optimize power flows between the PV plant and the grid; they are based on algorithms precisely conceived for maximizing the power injected into the grid by the PV plant and they are known as "active MPPTs".

In principle, these algorithms can be directly implemented on single-stage (DC-AC) PWM inverters; however, usually they are more profitably implemented on double-stage (DC-DC and DC-AC) PWM inverters, in which the DC-DC power electronic converter (widely, a boost type converter) is specifically devoted to the implementation of the selected MPPT algorithm.
There a lot of different MPPT algorithms, characterized by different performances, that are available for being implemented on PWM inverters; because this subject has been widely discussed and consolidated in the specialized literature [Esram and Chapman, 2007], here, these algorithms are not explicitly discussed.

3.2.2 Distributed active MPPTs

In case of mismatching conditions recalled in section 3.1.2, the use of a by-pass diode in parallel to each PV module of a PV string can mitigate losses of generating power but does not optimally solve the problem of maximizing the power generated by the PV string, neither by using, as interfacing apparatus, a PWM inverter endowed with a MPPT function.
In fact, for instance, please consider the presence of mismatched conditions in the case of a PV string constituted by two series-connected PV modules (each of one with a by-pass diode in parallel); the corresponding generation current-voltage characteristic curves of each PV module are depicted in Fig.2(a) [Shimizu et Al., 2003].
Furthermore, the characteristic curves of the output-current and of the output-power versus the output-voltage are as in Fig.2(b), that can be justified by means of the following considerations.
By varying the value of the equivalent resistance of the grid as seen by the PV string, the PV string current, Io, can goes from Ia to Id; the PV string voltage, Vo, and the PV string generated power, Po, change as follows:

- when Io = Ia (operating point A), both PV modules generate power, but neither one generates maximum power;
- when Io = Ib (operating point B), the shaded PV module generates maximum power, but the non-shaded PV module does not generate maximum power;
- when Io = Ic (operating point C), the non-shaded PV module generates power, but the shaded PV module does not generate any power, because the PV string current, Io, flows through its bypass diode;
- when Io = Id (operating point D), the non-shaded PV module generates its maximum power, but the shaded PV module does not generate any power because the PV string current, Io, flows through its bypass diode.

Finally, the output-power versus the output-voltage characteristic of the PV string is a multi-modal curve having a number of peaks (maximum) equal to the number of the string PV modules.
This means that, also with the use of the module by-pass diodes, in case of mismatching conditions the maximum output power of the PV string is less than the sum of the maximum generation power of the PV modules. In fact, with reference to the aforementioned situation, it is simple to underline that if the shaded PV module is not short-circuited, the non-shaded PV module does not operate at its maximum power; if the shaded PV module is short-circuited it does not generate its inherent power.
That said, it also evident that the presence of more than one maximum in the P-V characteristic of a PV string makes very difficult, for conventional active MPPTs, the detection of the absolute maximum power point of the PV string, to be caught.

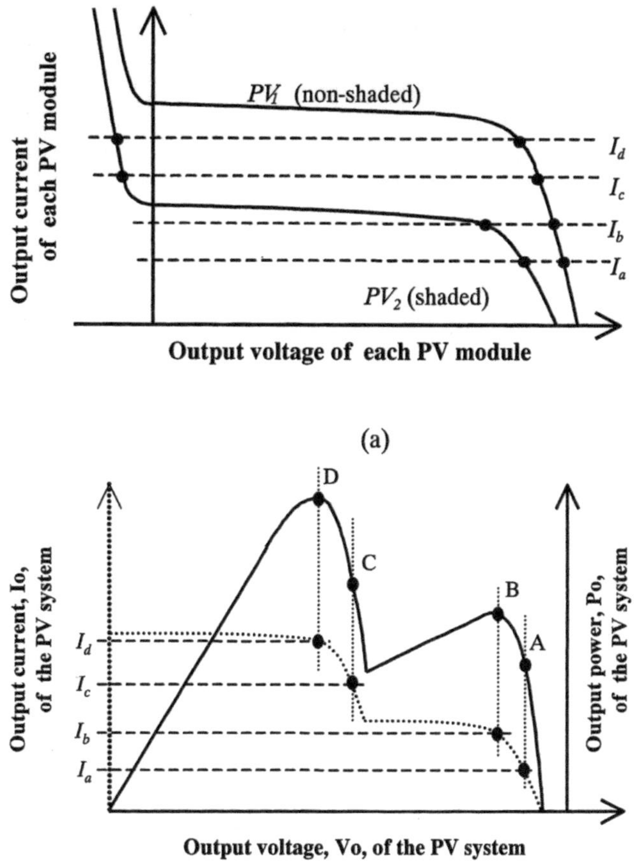

(b)

Fig. 2. PV module output current versus PV module output voltage, for different irradiation levels, 2(a), and effects of different solar irradiation levels on a string composed by two PV modules connected in series, 2(b). [Shimizu et Al., 2003]

Once analysed the presence of mismatching conditions in PV strings and their negative effects on their power generation, it is of practical relevance to investigate the possibility of finding a technical and cost-effective solution to such a problem, also taking into account of different layouts that are available for grid-connected PV plants and that have been discussed in a relevant number of recent papers and technical reports.

Intuitively, PV plant layouts with more inverters (string and/or module inverters) should be less sensitive to mismatching conditions with respect to PV plant with one centralized inverter only. However, in some recent studies [Woytea et al., 2003] it was also shown that this last assumption can not be always true and, in certain plant configurations, a plant layout with more inverters can result in partial shadowing related losses greater than that of a centralized inverter plant layout. This last situation is essentially related to the presence of multi-modal PV string or PV module power-voltage characteristics which create serious problems to conventional inverter active MPPT techniques. In fact, in presence of more than one local maximum points in the power-voltage characteristic of a PV string and/or a PV

module, conventional inverters implementing active MPPT techniques can catch and maintain one of the local maximum power points instead of the absolute maximum power point, so causing significant reductions in the PV plant whole generated power.
To overcome the aforementioned problems, essentially related to characteristics of conventional inverter active MPPT techniques and algorithms, in the recent specialized literature, different novel active MPPT techniques have been introduced and discussed, with the specific aim of maximizing the power generated by a PV string in case of significant mismatching conditions.
These new active techniques are essentially based on the idea that the MPPT function can be extended to each PV module of a string by implementing a so called distributed active MPPT (DMPPT) function.
Different techniques and algorithms have been, also, proposed in order to implement the aforementioned DMPPT. However, they are generally based on the use of expensive and complex high-frequency switching circuits and control techniques.
In [Shimizu et Al., 2003], a practical implementation of the aforementioned DMPPT technique is achieved by introducing a proper generation control circuit (GCC), that is to say a number of multi-stage DC/DC converter circuits to be connected in cascade to each module of a PV string, together with a centralized utility interactive inverter. In this way, the generation point of each PV module can be independently controlled and maximized. Furthermore, the P-V characteristic of a PV string results in a mono-modal characteristic with only a maximum power point to be caught by the MPPT algorithm, also in the presence of PV module mismatching conditions. A specific advantage of this technical solution is that power generation of each module can be maximized by physically detecting and controlling only the output power of the PV plants, so avoiding additional expensive power detectors to be installed on PV modules, on the contrary needed in other similar DMPPT techniques.
However, the proposed configuration circuit remains complex with respect to other circuit configurations with no distributed active DC/DC PV module converters.

4. Batteries as passive MPPTs

In this section it will be shown as batteries, connected in parallel to of PV field, can operate the aforementioned MMPT function passively, that is to say without the use of any additional circuitry and without implementing any control logic.
However, expected performances of this kind of passive MMPT technique are worse if compared with that of the aforementioned active techniques, mainly because of variation of the MPP caused by variations of working temperature of PV modules.
For these, reasons, after introducing the use of batteries in a centralized manner, this passive technique is extended to a different scenario in which batteries are contemplated to be more profitably used in a distributed manner.

4.1 The use of batteries in a centralized manner

For a fixed level of the solar irradiation and for a fixed temperature, the power delivered to the network by a conventional grid-connected PG, which qualitative I-V characteristic has been shown in Fig.1.(a), depends on the value of the network equivalent impedance as seen from the PG terminals, the behaviour of the network being normally imposed to be resistive by using a power electronic PWM inverters.

Unfortunately, only one working point on the I-V PG characteristic corresponds to the MPP (bold-faced circles, in Fig.1.(a) and it can be "caught" only carefully adjusting the value of the network equivalent resistance (i.e. by using the aforementioned active MPPT apparatus). On the other hand, Fig.1 also shows that, for different solar irradiation (SI) levels, the variation of the MPP directly involves the PG current values, while the PG voltage values remain almost constant.

This means that a battery, in parallel with the PG and the network as shown in Fig.1.(b), if properly designed in its nominal voltage value and in its capacity, can naturally catch and maintain a PG working point very close to the MPP, for any PG solar irradiation level and for any value of the network equivalent resistance (Fig.1.(b)).

Nevertheless, it is expected that battery used as passive MPPT has lower performances if compared with currently utilized active MPPTs, mainly because of the variations of the PG voltage at the MPP caused by variations of PG working temperature.

Because of this last consideration, as an alternative to the trivial idea of utilizing batteries in grid-connected PV systems in a centralized way and in the place of active MPPTs, it is proposed to use batteries in large grid-connected PV plants and in a more effective "distributed way", that is discussed in the next section.

4.2 The use of batteries in a distributed manner

The proposed idea of utilising batteries in large grid-connected PV plants and in a profitable "distributed way" can be explained and appreciated with the help of Fig.3.

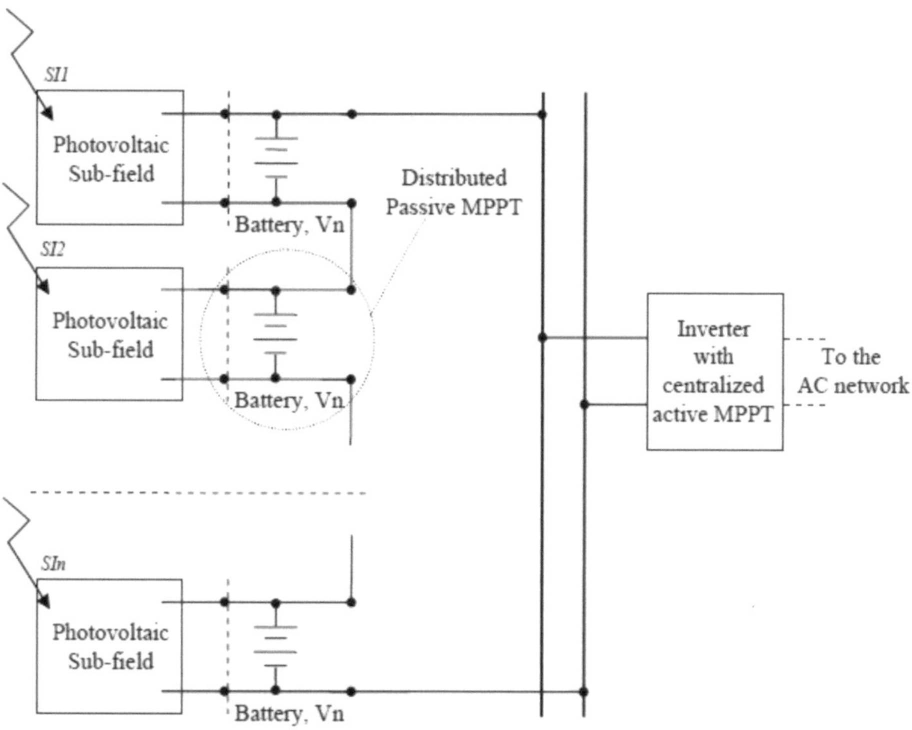

Fig. 3. Grid-connected photovoltaic system with distributed use of batteries as distributed passive MPPTs and centralized inverter with active MMPT function.

In practice, in large grid-connected PV plants, batteries can be used as passive and distributed MPPTs by locating them in parallel with a proper number of PV sub-fields and by maintaining a centralized inverters with centralized active MPPT function.
At the designing stage of the battery distributed system, there are some questions that have to be well analysed, in order to obtain an effective and inexpensive system. Firstly, the grouping of the PV modules, with which to build the PV sub-field. Then, the choice of the nominal voltage of the battery sub-systems to be connected in parallel to each PV sub-field. Finally, the choice of the capacity of the whole battery storage system.
The grouping of the PV modules is strictly connected to the choice of the nominal voltage of the battery sub-systems; it is worthwhile to remember that the nominal voltage of the battery sub-system has to be very close to the MPP voltage of the related PV sub-field and this last is approximately equal to 75-80% of the PV sub-field open circuit voltage.
In choosing the capacity of the battery system, the criterion to be followed is that of making a passive MPPT system that has to be effective but also inexpensive, in order to result as a valid alternative to the aforementioned, more complex and expensive, PV plant electrical schemes endowed by a number of distributed power electronic DC-DC converters. Furthermore, it is important to take into account also that: (i) energy generated by PV sub-fields - normally - has to be stored only for very short times (just to cope for the momentary unavailability of the grid); (ii) the active MPPT function of the centralized inverter should be able to avoid the full charge of batteries, even if they have a small capacity.

5. Experimental tests

In order to test the effectiveness of the proposed passive MPPT technique for grid-connected photovoltaic plants, that is based on the temporary storage of the generated power on a proper designed distributed battery system, two different experimental measurement campaigns have been performed; the first refers to laboratory experimental tests on a small-scale photovoltaic generator prototype of 20 Wp (section 5.1) while the second refers to experimental tests on a fully realized and installed grid-connected photovoltaic plant of 3 kWp (section 5.2).

5.1 Experiments on a small-power 20 Wp PV prototype

A small-scale prototype of about 20 Wp of power, constituted by 4 identical PV small-modules of about 5 Wp, together with 8 (2 for each panel) NiCd rechargeable batteries with 1.2 V of nominal voltage and 800 mAh of capacity, has been built-up and experimented under different artificial irradiation conditions (obtained by utilizing 4 flood light apparatus with 4x250 W lamps).
In practice, by following the guidelines of the electrical scheme depicted in Fig.3, in which the PV sub-fields are foreseen to be series connected, the prototype has been tested under both balanced and unbalanced artificial irradiation conditions.
In this experimental case study, the inverter is not present and, taking into account that in practice inverters for grid-connected PV plants have to inject into the grid a sinusoidal current in phase with the grid voltage, it has been substituted with a simple equivalent resistor, Ro; different values (optimal and non-optimal) for the inverter-grid equivalent resistance have been also considered in order to verify the effects of a conventional (centralized) MPPT function, typically performed by aforementioned inverters.

Even if a several number of experiments have been performed and analysed, for the sake of brevity, in the next, only the most interesting experimental results are referred and briefly discussed.

Specifically, in Tables I and II, the values of the power generated by each PV small-module and by the whole PV generator (4 series-connected small-modules), under balanced artificial irradiation levels and for optimal (Table I) and non optimal (Table II) values of the inverter-grid equivalent resistance, are summarized.

Experimental results confirm that, for a fixed and balanced (solar) irradiation (SI) level, a whole PV Plant can generate its inherent maximum power only if the inverter (thanks to its active MPPT function) is able to impose to the grid a resistive behaviour with a specific resistance value (optimal value).

In Tables III, the values of the power generated by each PV small-module and by the whole PV generator, under unbalanced artificial irradiation levels and for the optimal value of the inverter-grid equivalent resistance, are summarized.

Generated Power [W]		
	Without batteries	**With batteries**
PV Panel 1 (SI1 = 177 W/m^2)	0.63	0.61
PV Panel 2 (SI2 = 180 W/m^2)	0.64	0.62
PV Panel 3 (SI3 = 175 W/m^2)	0.63	0.60
PV Panel 4 (SI4 = 170 W/m^2)	0.61	0.59
Whole PV Plant	2.51	2.42

Table I. Results under balanced SI of PV Panels and optimal value of the inverter-grid equivalent resistance, Ro.

Generated Power [W]		
	Without batteries	**With batteries**
PV Panel 1 (SI1 = 177 W/m^2)	0.53	0.60
PV Panel 2 (SI2 = 180 W/m^2)	0.54	0.61
PV Panel 3 (SI3 = 175 W/m^2)	0.52	0.60
PV Panel 4 (SI4 = 170 W/m^2)	0.50	0.58
Whole PV Plant	2.09	2.39

Table II. Results under balanced SI of PV Panels and non optimal value of the inverter-grid equivalent resistance (-25% of Ro)

Generated Power [W]		
	Without batteries	With batteries
PV Panel 1 (SI1 = 224 W/m^2)	0.50	1.05
PV Panel 2 (SI2 = 220 W/m^2)	0.51	0.98
PV Panel 3 (SI3 = 66 W/m^2)	0.11	0.23
PV Panel 4 (SI4 = 280 W/m^2)	0.53	1.18
Whole PV Plant	1.65	**3.44**

Table III. Results under unbalanced SI of PV Panels and optimal value of the grid equivalent resistance, Ro

Experimental results show that, when the series connected PV sub-fields of a PV plant are irradiated in a non uniform way (unbalanced irradiation), even if in presence of a centralized inverter endowed with an active MPPT function, a conventional PV plants (without batteries) does not appear able to extract from the whole PV plant its inherent maximum power, that is to say the sum of the inherent maximum power of each PV sub-field; this is because the whole PV plant current, in a series connected PV sub-field system, is imposed to be equal to the lowest value among that of the different PV sub-fields.
Experimental results also show that, by introducing a properly designed distributed battery system, also in case of unbalanced solar irradiation conditions, it seems possible to significantly improve the power generated by the whole PV plant.

5.2 Experiments on a 3 kWp grid-connected PV plant

The analysis is specifically referred to a physically realized 18 kWp PV plant, connected to the LV distribution grid and owned by the B&T Solar Energy Division Company, on Lazzaro (RC), Italy. With some more details, the PV plant is realized by means of 90 PV modules with 200 Wp and with an open circuit voltage of about 40 V; PV modules are grouped in 3 PV sub-fields (with 30 modules and 6 kWp) connected to the grid by means of three different single-phase PWM inverters. Because of each inverter has two separate DC-inputs, with separate DC-DC boost-converter operating the MPPT function, each sub-field is realized with 2 different groups of 15 series-connected PV modules.
For our experimental tests and measurements, only one group of 15 series-connected PV modules (3 kWp) of a single-phase PV sub-field has been considered; furthermore, because of the temporarily limited availability of batteries, only 9 modules of the 15 series-connected PV modules have been experimented with (and without) batteries.
Specifically, 6 PV modules are always operated without batteries and are series-connected. On the other hand, the remaining 9 PV modules are grouped in 3 groups of 3 PV modules connected in parallel. Finally, the 3 parallel-connected PV modules are connected in series among them and also with the remaining 6 series-connected PV modules.
In order to test the effectiveness of the battery energy storage capability, a group of three 12V, 10 Ah, lead-acid series-connected batteries are connected (via a bipolar circuit breaker) in parallel to each aforementioned group of 3 parallel-connected PV modules.

Fig.4 should be helpful to graphically summarize the circuit configuration of the experimented PV plant.

Experimental tests have been operated under both balanced and unbalanced solar irradiation conditions of PV modules; in both cases, the results of measurements have been utilized to perform a comparative analysis between the power generated by the PV modules (and, then, by the whole PV sub-field), under the same irradiation conditions, without and with the use of batteries.

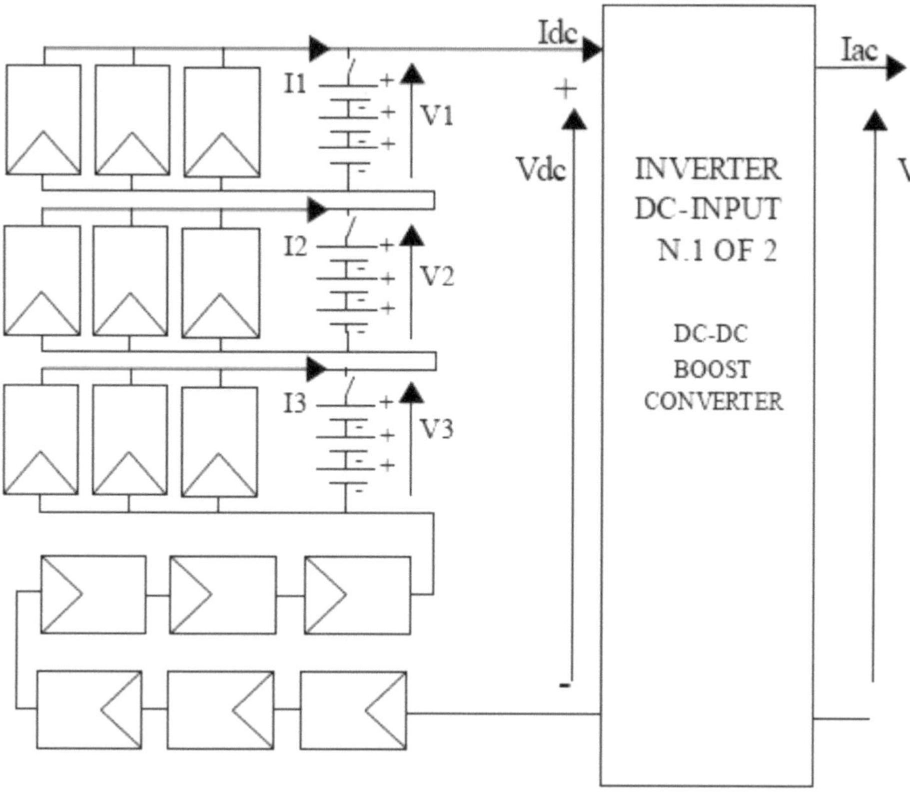

Fig. 4. Electrical scheme of the experimentally tested 3 kWp grid-connected PV plant

The first experimental test refers to balanced solar irradiation conditions. It is worthwhile to underline that, the weather being cloudy, the irradiation level was low (no more than 300 W/m^2). Measurements have been carried out for a total duration of one hour; during the first 30 minutes, batteries are switched-off while, during the second 30 minutes, batteries are switched-on. The results of measurements are summarised in Table IV in terms of registered time mean values. It is evident that, when connecting batteries, power generated by each group of PV modules, under the same solar irradiation conditions, increases. However, being the whole power of the PV plant injected into the grid, practically, the same than that injected without the presence of batteries, in this case the surplus of generated power is temporarily stored into the batteries and it will be injected into the grid later.

The second experimental test refers to unbalanced solar irradiation conditions; in practice, two PV modules have been artificially shadowed, as graphically evidenced in Fig.5.

Measured quantities *(time mean values)* - *symbols are as in Fig.4* -									
I1 [A]	**V1 [V]**	**I2 [A]**	**V2 [V]**	**I3 [A]**	**V3 [V]**	**Idc [A]**	**Vdc [V]**	**Iac [A]**	**Vac [V]**
Without batteries									
2,7	37,8	2,7	38,0	2,7	38,1	2,7	356,0	4,1	225,0
With batteries									
3,1	37,2	3,2	37,7	3,2	37,5	2,7	356,0	4,2	224,0

Table IV. Results of measurements under balanced solar irradiation conditions

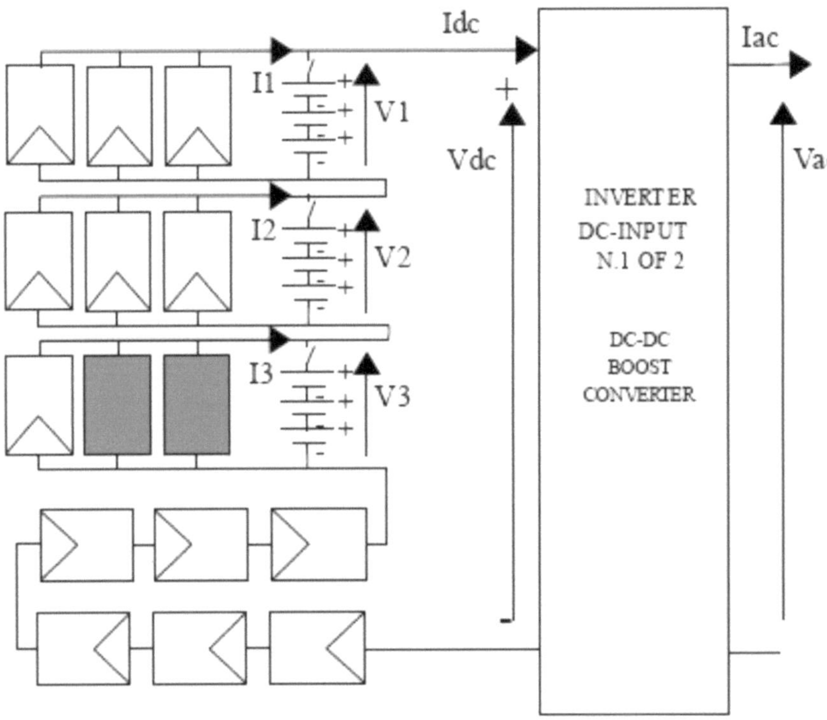

Fig. 5. Electrical scheme of the experimentally tested 3 kWp grid-connected PV plant, in presence of an artificially imposed partial shadowing of PV modules

Measurements have been carried out for a total duration of one hour:

- during the first interval of 15 minutes, batteries are switched-off and all the PV modules are irradiated,
- during the second interval of 15 minutes, batteries are switched-off and two PV modules are artificially shadowed;
- during the third interval of 15 minutes, batteries are switched-on and all the PV modules are irradiated,
- during the fourth interval of 15 minutes, batteries are switched-on and two PV modules are artificially shadowed.

The results are summarised in Table V in terms of registered time mean values.

Measured quantities *(time mean values)* *- symbols are as in Fig.5 -*									
I1 [A]	**V1 [V]**	**I2 [A]**	**V2 [V]**	**I3 [A]**	**V3 [V]**	**Idc [A]**	**Vdc [V]**	**Iac [A]**	**Vac [V]**
Without batteries and all PV-modules irradiated									
2,1	38,8	2,1	38,4	2,1	38,7	2,1	352,0	3,2	227,0
Without batteries and two PV-modules shadowed									
1,2	36,8	1,2	37,9	1,2	38,1	1,2	349,0	1,8	224,0
With batteries and all PV-modules irradiated									
1,9	37,2	2,2	37,7	2,0	37,7	2,3	347,0	3,6	223,0
With batteries and two PV-modules shadowed									
1,6	37	1,8	36,7	1,6	36,9	1,9	353,0	2,9	226,0

Table V. Results of measurements under unbalanced solar irradiation conditions

In order to better summarize the experimental results and to better appreciate what happens, especially in case of unbalanced solar irradiation conditions, with and without the use of batteries, let us to define a Power Decay Coefficient, PDC%, that is to say a coefficient that measures the decay rate of the power generated by all the PV modules as a consequence of the shadowing of only some PV modules of the whole PV field:

$$PDC\,\% = \frac{(\text{Power without shadows - Power with shadows})}{\text{Power without shadows}} x100 \qquad (1)$$

Then, with reference to Table V, the value of PDC%, without and with batteries, are reported in Fig.6.

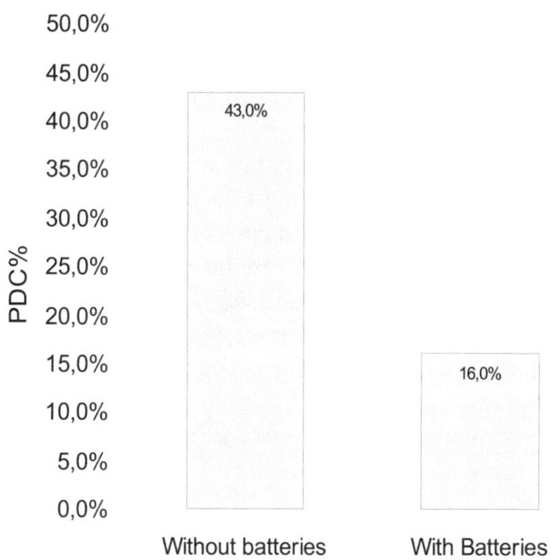

Fig. 6. Decay rate (PDC%) of the PV plant generated power, in presence of partial shadowing of PV modules, without ant with the use of batteries, as suggested in Fig.3.

It is possible to note that, partial shadowing of only a limited number of PV modules, in conventional (without batteries) PV plants, can cause an important decay of the whole generated power (43%); on the contrary, in presence of batteries, the same partial shadowing causes a decay rate of the whole generated power significantly lower (only 16%).

6. Perspectives on developing single AC PV modules, with on board distributed batteries used for energy storage

From results and considerations of previous sections, it can be summarized that the intrinsic variability of solar irradiation forces conventional grid-connected PV plants to inject power into the grid in a way as variable and unpredictable.

Furthermore, as well known, conventional PWM inverters, for connecting PV plants to distribution grids, generate an AC output voltage characterized by harmonic and inter-harmonic components (especially at high frequencies, in the range of their switching function). Even if output filters are conventionally used, remaining harmonics and inter-harmonics may cause different power quality problems, especially in terms of malfunctioning of information and communication technology (ICT) apparatus, that are more and more utilized in modern distribution grids.

Currently, in the specialized scientific literature, researchers are brightly discussing about the possibilities to develop new power electronic apparatus for interconnecting PV plants and the distribution grids with power quality problems reduced with respect to that caused by conventional PWM inverters [Busquets-Monge et al., 2008].

On this basis, on the opinion of the Author, the idea here investigated to introduce in grid-connected PV plants an energy storage system, based on a conspicuous number of batteries with small capacity and operated in a distributed manner, can be utilized also for defining and developing new multi-level power electronic inverters [Khomfoi & Tolbert, 2007] intrinsically characterized by AC output voltages with very high quality waveforms and, also, by high reliability and availability.

Particularly interesting could be the idea of developing single PV modules able to generate an AC output voltage (AC PV modules) directly compatible with the low voltage distribution grids and with high quality waveform, being this achievable by means of a proper designed and developed multi-level inverter installed on the PV modules.

With some more details, by installing on a conventional PV module a conspicuous number of small rechargeable batteries, to be put in parallel to a proper group of series connected PV cells, an as many conspicuous number of DC voltage levels is physically available on board of the PV module and these DC voltage levels can be utilized, by a proper designed and developed multi-level electronic inverter, to build up an AC quasi-sinusoidal voltage at the distribution grid frequency; an isolation transformer (a HF-transformer on the DC section of the circuit or a LF-transformer on the AC output section of the circuit) could be also utilized to adjust the AC output voltage rms value of the PV module and to cope for galvanic isolation.

In addition to the high quality waveform of the AC output voltage, batteries installed on the PV module would make it more efficient, available and reliable.

7. Conclusion

A passive MPPT technique, to be utilized mostly in large grid-connected PV plants, has been introduced and discussed; it is essentially based on the energy storage capabilities of batteries that are proposed to be put in parallel to a proper number of PV sub-fields, so as to be used in a distributed manner. If well designed in their location, in their nominal voltage value and in their capacity, batteries can naturally catch the MPP of each PV sub-field, also compensating for critical unbalanced solar irradiation conditions.

The results of different experimental tests, operated both on a very small-power 20 Wp prototype and on a 3 kWp physically realized grid-connected PV plant, have clearly demonstrated the effectiveness of the proposed technique, also showing that, in some critical irradiation conditions, batteries used in grid-connected PV plants can significantly increase the energy generation with respect to that of a conventional PV plant. The proposal can be a valid and lower cost alternative to more expensive solutions based on a number of DC-DC power electronic converters to be put in parallel to each PV sub-field in order to work as distributed active MPPTs.

Furthermore, the presence of an energy storage system can make more and more attractive grid-connected PV plants, due to some important additional capabilities not commons of currently conceived grid-connected PV plants, as: a more great availability in favour of the AC power grid; a significant reduction of unfavourable requests of occasional peaks of load power demand; the possibility to substitute other expensive (and often not renouncing) apparatus for utility grid power quality improvements, as UPS and active filters; the possibility to be integrable with other different renewable resources, with minor expenses and with great economical advantages.

Finally, a conspicuous number of batteries distributed on board to a single PV module could be on the basis of the development of AC PV modules, to be directly connected to LV distribution grids and characterized by high quality of AC voltage, high efficiency and high availability.

8. References

Busquets-Monge, S.; Rocabert, J.; Rodriguez, P.; Alepuz, S.; Bordonau, J. (2008). Multilevel Diode clamped Converter for Photovoltaic Generators with Independent Voltage Control of Each Solar Array. *IEEE Transactions on Industrial Electronics*, Vol.55, July 2008, pp. 2713-2723.

Carbone, R. (2009). Grid-Connected Photovoltaic Systems with Energy Storage. *Proceeding of IEEE International Conference on CLEAN ELECTRICAL POWER, Renewable, Energy Resources Impact "ICCEP 2009"*. Capri - Italy, June 9-11, 2009.

Denholm, Paul; Margolis, Robert M. (2007). Evaluating the limits of solar photovoltaics (PV) in electric power systems utilizing energy storage and other enabling technologies. *ELSEVIER, Energy Policy 35*, (2007) 4424–4433. www.elsevier.com/locate/enpol.

Esram, T.; Chapman, P.L. (2007). Comparison of Photovoltaic Array Maximum Power Point Tracking Techniques. *IEEE Transactions on Energy Conversion*, Vol.22, N.2, June, 2007, pp.439-449.

Khomfoi, S.; Tolbert, L.M. (2007). Multilevel Power Converters. *Power Electronics Handbook, 2nd Edition, Elsevier*, 2007, ISBN 978-0-12-088479-7, Chapter 17, pp. 451-482.

Lo, Y. K.; Lin, J. Y.; Wu, T. Y. (2005). Grid-Connection Technique for a Photovoltaic System with Power Factor Correction. *IEEE PEDS 2005*, 0-7803-9296-5/05.

Lu, B.; Shahidehpour, M. (2005). Short-Term Scheduling of Battery in a Grid-Connected PV/Battery System. *IEEE Transactions on Power Systems*, Vol. 20, N°2, May 2005.

Nourai, Ali; Kearns, David. (2010). Realizing Smart Grid Goals with Intelligent Energy Storage. *IEEE power & energy magazine*, Vol. 49, march/april 2010. 657-R114.

Shimizu,T.; Hashimoto, O.; Kimura, G. (2003). A Novel High-Performance Utility-Interactive Photovoltaic Inverter System. *IEEE Transactions on Power Electronics*, Vol.18, 2003, N. 2.

Ueda, Yuzuru; Kurokawa, Kosuke; Itou, Takamitsu; Kitamura, Kiyoyuki; Akanuma, Katsumi; Yokota, Masaharu; Sugihara, Hiroyuki; Morimoto, Atsushi. (2006). Performance analyses of battery integrated grid-connected residential PV systems. *Proceeding of 21st European Photovoltaic Solar Energy Conference*, 4-8 September 2006, Dresden, Germany.

Woytea, Achim; Nijsa, Johan; Belmans, Ronnie. (2003). Partial shadowing of photovoltaic arrays with different system configurations: literature review and field test results. *ELSEVIER Transaction on Solar Energy*, Vol. 74, 2003, pp. 217–233.

5

Multi-Area Frequency and Tie-Line Power Flow Control by Fuzzy Gain Scheduled SMES

M.R.I. Sheikh[1], S.M. Muyeen[2], R. Takahashi[3], and J. Tamura[3]
[1]EEE Department, Rajshahi University of Engineering & Technology, Rajshahi – 6204,
[2]EEE Department, The Petroleum Institute, P.O.: 2533, Abu Dhabi,
[3]EEE Department, Kitami Institute of Technology, 165 Koen-cho, Kitami, 090-8507,
[1]Bangladesh
[2]U.A.E
[3]Japan

1. Introduction

Generation and distribution of electric energy with good reliability and quality is very important in power system operation and control. This is achieved by Automatic Generation Control (AGC). In an interconnected power system, as the load demand varies randomly, the area frequency and tie-line power interchange also vary. The objective of Load Frequency Control (LFC) is to minimize the transient deviations in these variables and to ensure for their steady state values to be zero. The LFC performed by only a governor control imposes a limit on the degree to which the deviations in frequency and tie-line power exchange can be minimized. However, as the LFC is fundamentally for the problem of an instantaneous mismatch between the generation and demand of active power, the incorporation of a fast-acting energy storage device in the power system can improve the performance under such conditions. But fixed gain controllers based on classical control theories are presently used. These are insufficient because of changes in operating points during a daily cycle [Benjamin et al., 1978; Nanda et al., 1988; Das et al., 1990; Mufti et al., 2007; Nanda et al., 2006 & Oysal et al., 2004] and are not suitable for all operating conditions. Therefore, variable structure controller [Benjamin et al., 1982; Sivaramaksishana et al., 1984; Tripathy et al., 1997 & Shayeghi et al., 2004] has been proposed for AGC. For designing controllers based on these techniques, the perfect model is required which has to track the state variables and satisfy system constraints. Therefore it is difficult to apply these adaptive control techniques to AGC in practical implementations. In multi-area power system, if a load variation occurs at any one of the areas in the system, the frequency related with this area is affected first and then that of other areas are also affected from this perturbation through tie lines. When a small load disturbance occurs, power system frequency oscillations continue for a long duration, even in the case with optimized gain of integral controllers [Sheikh et al., 2008 & Demiroren, 2002]. So, to damp out the oscillations in the shortest possible time, automatic generation control including SMES unit is proposed.
Therefore, in the proposed control system, with an addition of the simple SMES controller, a supplementary controller with K_{Ii} (as shown in Fig. 6) is designed in order to retain the

frequency to the set value after load changes. These controllers must eliminate the frequency transients as soon as possible. Using fuzzy logic, the integrator gain (K_{Ii}) of the supplementary controller is so scheduled that it compromise between fast transient recovery and low overshoot in dynamic response of the system. It is seen that with the addition of gain scheduled supplementary controller, a simple controller scheme for SMES is sufficient for load frequency control of multi-area power system [Sheikh et al., 2008].

2. Superconducting Magnetic Energy Storage (SMES) system

2.1 Overview of SMES

A superconducting magnetic energy storage system is a DC current device for storing and instantaneously discharging large quantities of power. The DC current flowing through a superconducting wire in a large magnet creates the magnetic field. The large superconducting coil is contained in a cryostat or dewar consisting of a vacuum vessel and a liquid vessel that cools the coil. A cryogenic system and the power conversion/conditioning system with control and protection functions [IEEE Task Force, 2006] are also used to keep the temperature well below the critical temperature of the superconductor. During SMES operation, the magnet coils have to remain in the superconducting status. A refrigerator in the cryogenic system maintains the required temperature for proper superconducting operation. A bypass switch is used to reduce energy losses when the coil is on standby. And it also serves other purposes such as bypassing DC coil current if utility tie is lost, removing converter from service, or protecting the coil if cooling is lost [M. H. Ali et al., 2008].
Figure 1 shows a basic schematic of an SMES system [http://www.doc.ic.ac.uk/~matti/ise 2grp/energystorage_report/node8.html]. Utility system feeds the power to the power conditioning and switching devices that provides energy to charge the coil, thus storing energy. When a voltage sag or momentary power outage occurs, the coil discharges through switching and conditioning devices, feeding conditioned power to the load. The cryogenic (refrigeration) system and helium vessel keep the conductor cold in order to maintain the coil in the superconducting state.

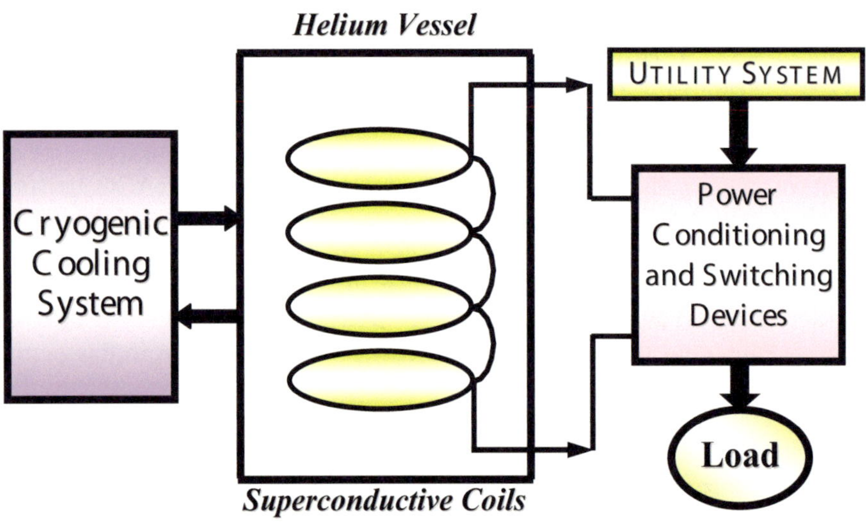

Fig. 1. Schematic diagram of the basic SMES system

2.2 Advantages of SMES

There are several reasons for using superconducting magnetic energy storage instead of other energy storage methods. The most important advantages of SMES are that the time delay during charge and discharge is quite short. Power is available almost instantaneously and very high power output can be provided for a brief period of time. Other energy storage methods, such as pumped hydro or compressed air have a substantial time delay associated with the conversion of stored mechanical energy back into electricity. Thus if a customer's demand is immediate, SMES is a viable option. Another advantage is that the loss of power is less than other storage methods because the current encounters almost zero resistance. Additionally the main parts in a SMES are motionless, which results in high reliability. Also, SMES systems are environmentally friendly because superconductivity does not produce a chemical reaction. In addition, there are no toxins produced in the process.

The SMES is highly efficient at storing electricity (greater than 97% efficiency), and provide both real and reactive power. These systems have been in use for several years to improve industrial power quality and to provide a premium-quality service for individual customers vulnerable to voltage and power fluctuations. The SMES recharges within minutes and can repeat the charge/discharge sequence thousands of times without any degradation of the magnet [http://en.wikipedia.org/wiki/Superconducting_magnetic_energy_storage]. Thus it can help to minimize the frequency deviations due to load variations [Demiroren & Yesil, 2004]. However, the SMES is still an expensive device.

2.3 SMES for Load Frequency Control application

A sudden application of a load results in an instantaneous mismatch between the demand and supply of electrical power because the generating plants are unable to change the inputs to the prime movers instantaneously. The immediate energy requirement is met by the kinetic energy of the generator rotor and speed falls. So system frequency changes though it becomes normal after a short period due to Automatic Generation Control. Again, sudden load rejections give rise to similar problems. The instantaneous surplus generation created by removal of load is absorbed in the kinetic energy of the generator rotors and the frequency changes. The problem of minimizing the deviation of frequency from normal value under such circumstances is known as the load frequency control problem. To be effective in load frequency control application, the energy storage system should be fast acting i.e. the time lag in switching from receiving (charging) mode to delivering (discharging) mode should be very small. For damping the swing caused by small load perturbations the storage units for LFC application need to have only a small quantity of stored energy, though its power rating has to be high, since the stored energy has to be delivered within a short span of time. However, due to high cost of superconductor technology, one can consider the use of non-superconducting of lossy magnetic energy storage (MES) inductors for the same purpose. Such systems would be economical maintenance free, long lasting and as reliable as ordinary power transformers.

Thus a MES system seems to be good to meet the above requirements. The power flow into an energy storage unit can be reversed, by reversing the DC voltage applied to the inductor within a few cycles. A 12-pulse bridge converter with an appropriate control of the firing angles can be adopted for the purpose. Thus, these fast acting energy storage devices can be made to share the sudden load requirement with the generator rotors, by continuously controlling the power flow in or out of the inductor depending on the frequency error signals.

3. Analysis of the magnetic energy storage unit

The SMES inductor converter unit for improvement in power system LFC application essentially consists of a DC inductor, an ac/dc converter and a step down Y-Y/Δ transformer. The inductor should be wound with low resistance, large cross-section copper conductors. The converter is of the 12-pulse cascaded bridge type shown in Fig. 2, connected to the inductor in the DC side and to the three-phase power system bus through the transformer in the ac side [R.J. Abraham et al., 2008]. Control of the firing angles of the converter enables the DC voltage applied (V_{sm}) to the inductor to be varied through a wide range of positive and negative values as shown in Fig. 3. Gate turn off thyristors (GTO)

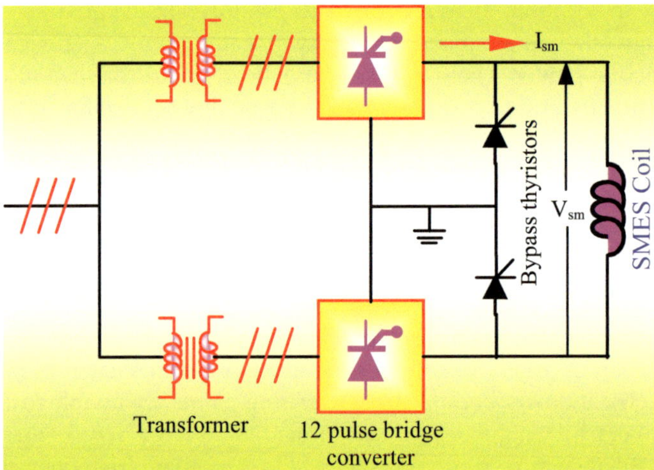

Fig. 2. Schematic diagram of the SMES unit

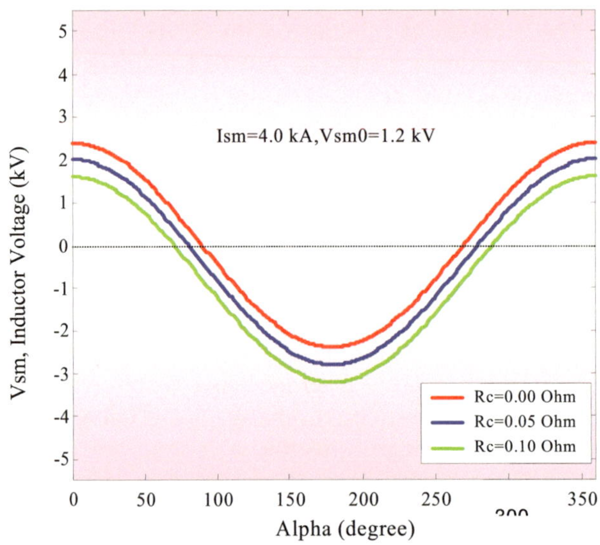

Fig. 3. Effect of inductor voltage, V_{sm} with the variation of firing angle of 12-pulse converter

allow us to design such type of converter. When charging the magnet, a positive DC voltage is applied to the inductor. The current in the inductor rises exponentially or linearly and the magnetic energy is stored. When the current reaches the rated value, the applied voltage is brought down to low value, sufficient to overcome the voltage drop due to inductor resistance. When the extra energy is required in the power system, a negative DC voltage is applied to the inductor by controlling the firing angles of the converter. The losses in the MES unit would consist of the transformer losses, the converter losses, and the resistive loss in the inductor coil. The inductor loss can be kept at an acceptable level by proper design of the winding.

Due to sudden application or rejection of load, the generator speed fluctuates. When the system load increases, the speed falls at the first instant. However, due to the governor action, the speed oscillates around some reference value. The converter works as an inverter ($90^\circ < |\alpha| < 270^\circ$) when the actual speed is less than the reference speed and energy is withdrawn from the SMES unit (P_{sm} negative). However, the energy is recovered when the speed swings to the other side.

The converter then works as a rectifier ($-90^\circ < |\alpha| < 90^\circ$) and the power P_{sm} becomes positive.

If the transformer and converter losses are neglected, according to the circuit analysis of converter, the voltage V_{sm} of the D.C side of the 12-pulse converter under equal-α (EA, when $\alpha_1 = \alpha_2 = \alpha$) mode is expressed by

$$V_{sm} = V_{sm0}(\cos\alpha_1 + \cos\alpha_2) = 2\,V_{sm0}\cos\alpha - 2\,I_{sm}R_c \tag{1}$$

where

α is the firing angle

V_{sm} is the DC voltage applied to the inductor

I_{sm} is the current through the inductor

R_c is the equivalent commutating resistance and

V_{sm0} is the maximum open circuit bridge voltage of each 6-pulse bridge at $\alpha=0$.

When the inductor is charged initially, the current build up, expressed, as a function of time with V_{sm} held constant, is given as

$$I_{sm} = \frac{V_{sm}}{R_L}\left(1 - e^{-\frac{R_L}{L}t}\right) \tag{2}$$

where L and R_L are the inductance and the resistance of inductor respectively.

Once the current reaches its rated value I_{sm0} it is held constant by reducing the voltage to a value V_{sm0} enough to overcome the resistive drop. In this case

$$V_{sm0} = I_{sm0}.\,R_L \tag{3}$$

As this value of V_{sm0} is very small, the firing angle will be nearly 90^0. At any instant of time the amount of energy stored in the inductor is given by

$$W_{sm} = W_{sm0} + \int_{t_0}^{t} P_{sm}\,d\tau \tag{4}$$

where, $W_{sm0} = \frac{1}{2}LI_{sm0}^2$ is the initial energy in the inductor.

Once the rated current in the inductor is reached, the unit is ready to be coupled with the power system application. The frequency deviation Δf of the power system is sensed and fed to the MES unit as the error signal. ΔV_{sm} is then continuously controlled depending on this signal. When there is a sudden increase in load in the power system, the frequency falls and a negative voltage, expressed by equation

$$\Delta V_{sm} = K_0 \, \Delta f \tag{5}$$

is impressed on the inductor. The converter bridges maintain a unidirectional current flow and as the circuit is inductive the current does not change instantaneously. In this mode of operation, a positive converter voltage produces positive power, which means charging the coil, and a negative converter voltage produces a negative power and discharges the inductor. When the frequency dip in the power system causes a negative voltage to be applied to the inductor, power flows from the inductor into the power system, sharing the sudden load requirement. The reverse process takes place when there is a sudden load rejection in the power system. The frequency increase causes a positive voltage to be impressed on the inductor and the MES unit absorbs the excess power from the power system. The conceptual diagram of active and reactive power modulation under equal-α mode is shown in Fig. 4.

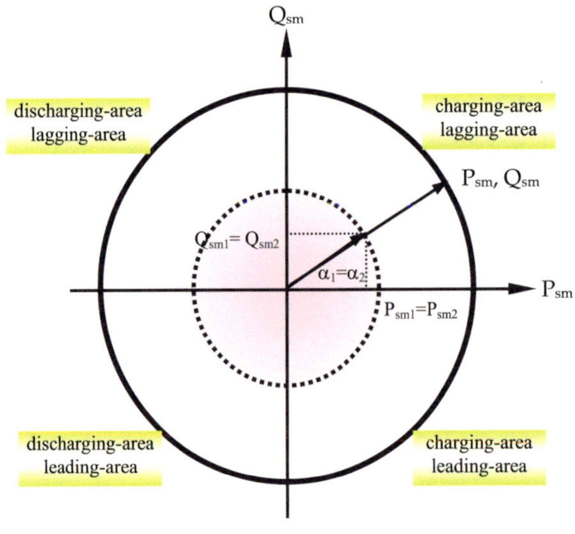

Fig. 4. Conceptual diagram of active and reactive power control of MES unit under equal-α mode

In actual practice the inductor current should not be allowed to reach zero to prevent the possibility of discontinuous conduction in the presence of the large disturbances [Wu et al., 1991 & Banerjee et al., 1990]. It is desirable to set the rated inductor current I_{sm0} such that the maximum allowable energy absorption equals the maximum allowable energy discharge [Wu et al., 1991 & Banerjee et al., 1990]. This makes the SMES equally effective in damping swings caused by sudden increase as well as decrease in load. Thus, if the lower current limit is chosen at 0.3 I_{sm0}, the upper inductor current limit, based on the equal energy absorption/discharge criterion becomes 1.38 I_{sm0} [Banerjee et al., 1990]. When the inductor current reaches either of these limits, the dc voltage has to be brought to zero.

As the inductor has a finite inductance and hence a finite amount of energy stored in it, the current in the inductor falls as energy is withdrawn from the coil. This deviation in the inductor current is expressed as

$$\Delta I_{sm} = \frac{\Delta V_{sm}}{R_L + s.L} \tag{6}$$

Prior to the load disturbance, let the magnitudes of voltage and current are V_{sm0} and I_{sm0} (nominal values). Thus the initial power flow into the coil can be expressed as

$$P_{sm0} = V_{sm0} \,.\, I_{sm0} \tag{7}$$

In response to the load disturbance the incremental change of power flow into the coil can be expressed as

$$\Delta P_{sm} = I_{sm0}.\, \Delta V_{sm} + I_{sm0}.\, R_L.\, \Delta I_{sm} + \Delta V_{sm}\,.\Delta I_{sm} \tag{8}$$

Following a sudden increase in load in the power system, the incremental power expressed by equation (8) is discharged into the power system by the energy storage unit to share with the generator rotor, the extra load demand.

4. Integration of SMES with two-area power system

Figure 5 shows the proposed configuration of SMES units in a two-area power system [Mufti et al., 2007]. Two areas are connected by a weak tie-line. When there is sudden rise in power demand in a control area, the stored energy is almost immediately released by the SMES through its power conversion system (PCS). As the governor control mechanism starts working to set the power system to the new equilibrium condition, the SMES coil stores energy back to its nominal level. Similar action happens when there is a sudden decrease in load demand. Basically, the operation speed of governor-turbine system is slow compared with that of the excitation system. As a result, fluctuations in terminal voltage can be corrected by the excitation system very quickly, but fluctuations in generated power or frequency are corrected slowly. Since load frequency control is primarily concerned with the real power/frequency behavior, the excitation system model will not be required in the approximated analysis [Mufti et al., 2007 & Sheikh et al., 2008]. This important simplification paves the way for constructing the simulation model shown in Fig. 6.

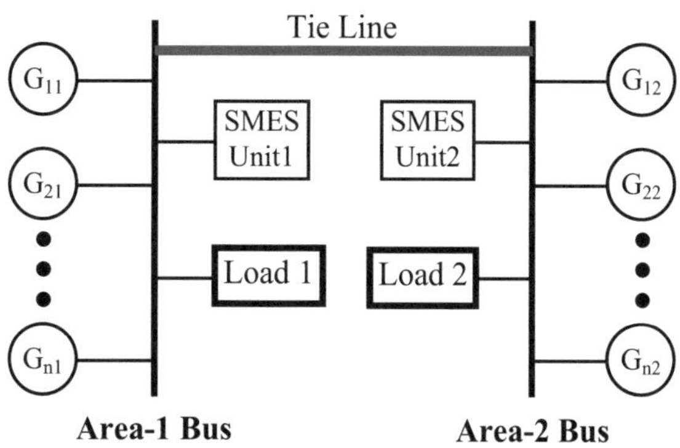

Fig. 5. Configuration of SMES in a two-area power system

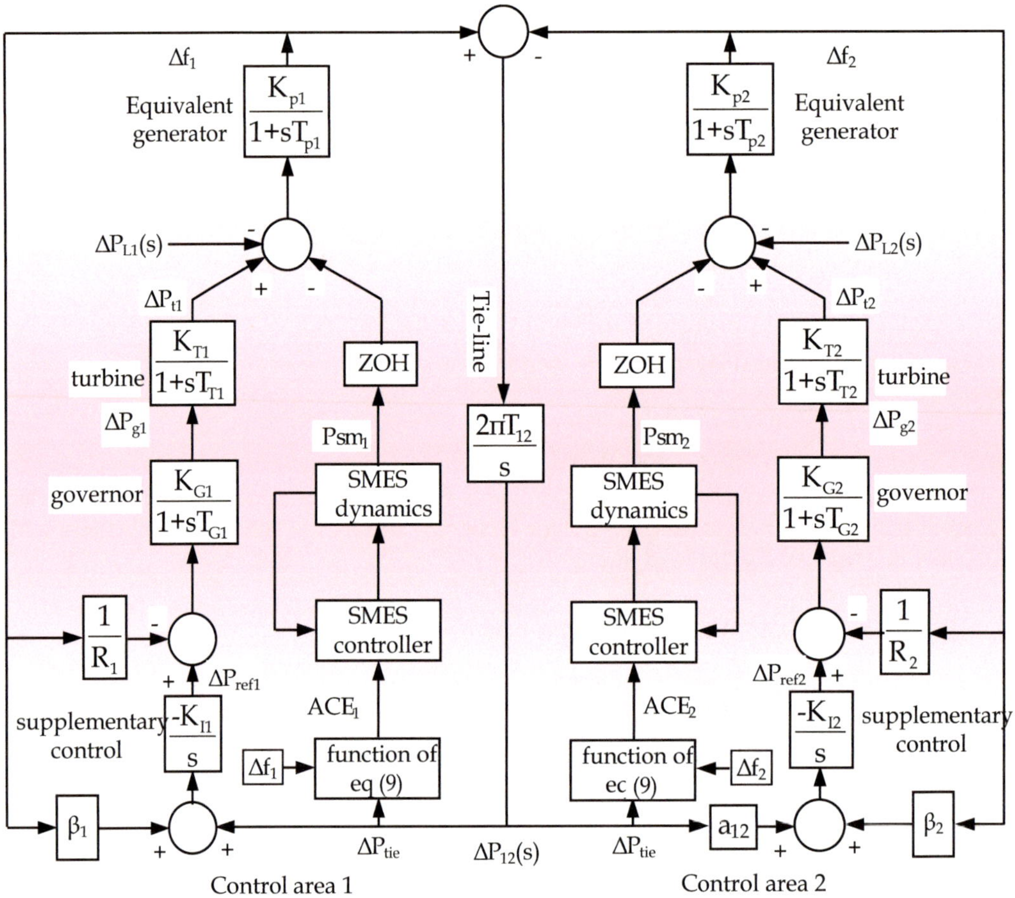

Fig. 6. Digital simulation model for the two-area power system

The basic objective of the supplementary control is to restore balance between each area load and generation for a load disturbance. This is met when the control action maintains the frequency and the tie-line power interchange at the scheduled values. The supplementary controller with integral gain K_{Ii} is therefore made to act on area control error (ACE), which is a signal obtaind from tie-line power flow deviation added to frequency deviation weighted by a bias factor β.

$$ACE_i = \sum_{j=1}^{n} \Delta P_{tie,\, i\, j} + \beta_i \Delta f_i \tag{9}$$

where the suffix i refer to the control area and j refer to the number of generator. All parameters are same as those used in [Sheikh et al., 2008]

5. Optimization of integral gain, K_I and frequency bias factors, β

Figure 7 shows the frequency deviations for different values of K_I for a specific load change. It is observed that a higher value of K_I results in reduction of maximum deviation of the system frequency but the system oscillates for longer times. Decreasing the value of K_I yields comparatively higher maximum frequency deviation at the beginning but provides very good damping in the later cycles. These initiate a variable K_I, which can be determined from the frequency error and its derivative. Obviously, higher values of K_I is needed at the initial stage and then it should be changed gradually depending on the system frequency changes.

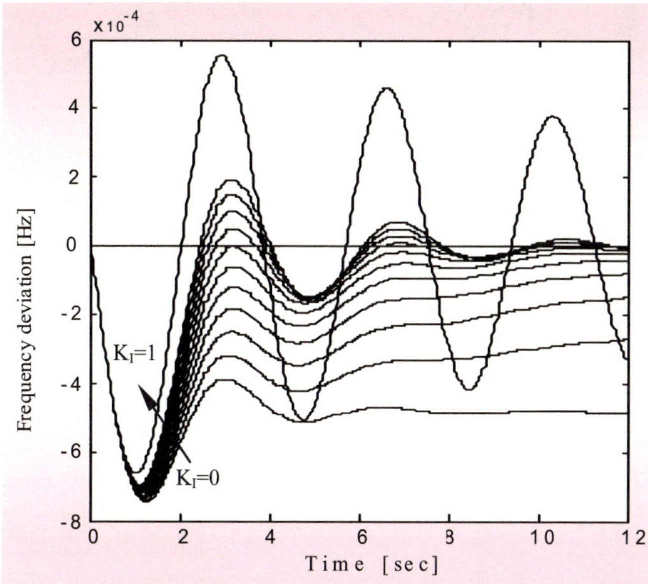

Fig. 7. Frequency deviation step response for different values of K_I

Dynamic performance of the AGC system would obviously depend on the value of frequency bias factors, $\beta_1 = \beta_2 = B$ and integral controller gain value, $K_{I1}=K_{I2}=K_I$. In order to optimize B and K_I the concept of maximum stability margin is used, evaluated by the eigen-values of the closed loop control system.

For a fixed gain supplementary controller, the optimal values of K_I and B are chosen, here, on the basis of a performance index (PI) given in (10) for a specific load change. The Performance Index (PI) curves are shown in Fig. 8 without considering governor dead-band (DB) and generation rate constraints (GRC).

$$PI = \int_0^T \left(\Delta P_{tie}^2 + w_1 \Delta f_1{}^2 + w_2 \Delta f_2^2\right) dt \tag{10}$$

Where, w_1 and w_2 are the weight factors. The weight factors w_1 and w_2 both are chosen as 0.25 for the system under consideration [Sheikh et al., 2008].

From Fig. 8, in the absence of DB & GRC it is observed that the value of integral controller gain, $K_I = 0.34$ and frequency bias factors, B=0.4 which occurs at PI = 0.009888.

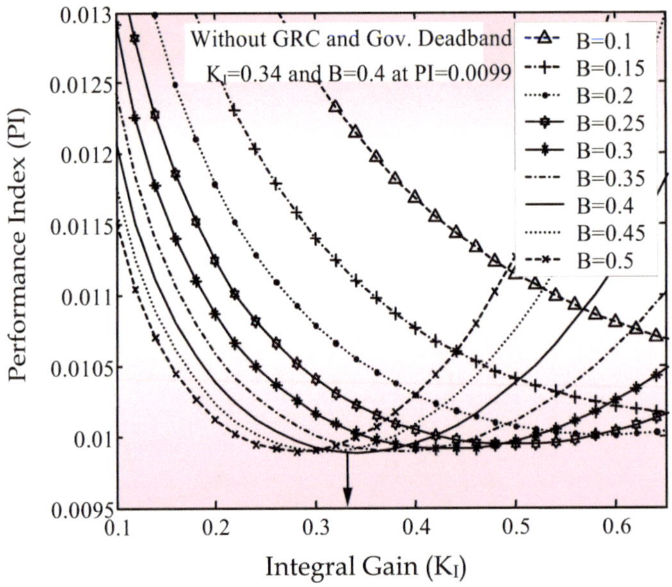

Fig. 8. The optimal integral controller gain, K_I and frequency bias factor, B

6. Control system design

6.1 Fuzzy gain schedule PI controller for AGC [Sheikh et al., 2008]

Figure 9 shows the membership functions for PI control system with a fuzzy gain scheduler. The approach taken here is to exploit fuzzy rules and reasoning to generate controller parameters. The triangular membership functions for the proposed fuzzy gain scheduled integral (FGSPI) controller of the three variables (e_t, $c\dot{e}_t$, K_I) are shown in Fig. 9, where frequency error (e_t) and change of frequency error ($c\dot{e}_t$) are used as the inputs of the fuzzy logic controller. K_{Ii} is the output of fuzzy logic controller. Considering these two inputs, the output of gain K_{Ii} is determined. The use of two input and single output variables makes the design of the controller very straightforward. A membership value for the various linguistic variables is calculated by the rule given by

$$\mu\left(e_t,c\dot{e}_t\right)=\min\left[\mu\left(e_t\right),\mu\left(c\dot{e}_t\right)\right] \tag{11}$$

The equation of the triangular membership function used to determine the grade of membership values in this work is as follows:

$$A\left(x\right)=\frac{\left(b-2\left|x-a\right|\right)}{b} \tag{12}$$

Where A(x) is the value of grade of membership, 'b' is the width and 'a' is the coordinate of the point at which the grade of membership is 1 and 'x ' is the value of the input variables. The control rules for the proposed strategy are very straightforward and have been developed from the viewpoint of practical system operation and by trial and error methods.

The membership functions, knowledge base and method of defuzzification determine the performance of the FGSPI controller in a multi-area power system as shown in (13). Mamdani's max-min method is used. The center of gravity method is used for difuzzification to obtain K_I. The entire rule base for the FGSPI controller is shown in Table I.

$$K_I = \frac{\sum_{j=1}^{n} \mu_j u_j}{\sum_{j=1}^{n} \mu_j} \tag{13}$$

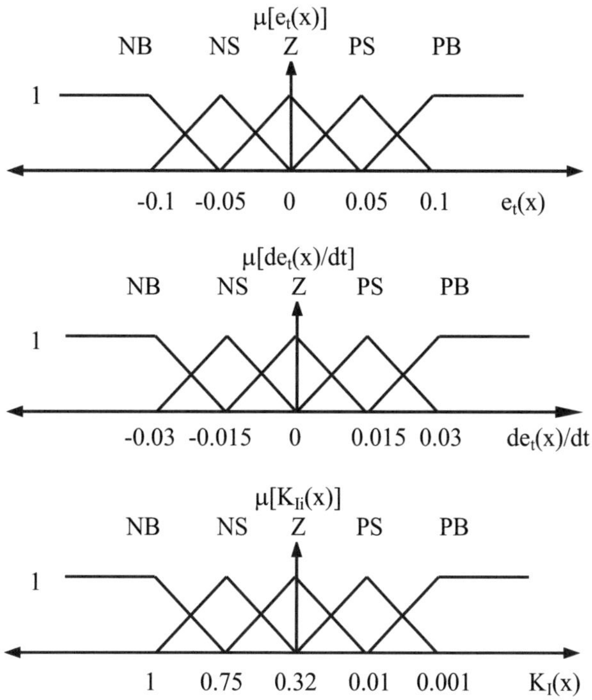

Fig. 9. Membership functions for the fuzzy variables

ce \ e	NB	NS	Z	PS	PB
NB	PB	PB	PB	PS	Z
NS	PB	PB	PS	Z	NS
Z	PB	PS	Z	NS	NB
PS	PS	Z	NS	NB	NB
PB	Z	NS	NB	NB	NB

Table 1. Fuzzy Rule base for FGSPI Controller

6.2 Control strategy for SMES

Figure 10 outlines the proposed simple control scheme for SMES, which is incorporated in each control area to reduce the instantaneous mismatch between the demand and generation, where I_{sm}, V_{sm} and P_{sm} are SMES current, SMES voltage and SMES power respectively. For operating point change due to load changes, gain (K_{Ii}) scheduled supplementary controller is proposed. Firstly K_{Ii} is determined using the fuzzy controller to obtain frequency deviation, Δf, and tie-line power deviation, ΔPtie. Finally ACE_i which is the combination of ΔPtie and Δf [as shown in (9)] is used as the input to the SMES controller. It is desirable to restore the inductor current to its rated value as quickly as possible after a system disturbance, so that the SMES unit can respond properly to any subsequent disturbance. So inductor current deviation is sensed and used as negative feedback signal in the SMES control loop to achieve quick restoration of current and SMES energy levels.

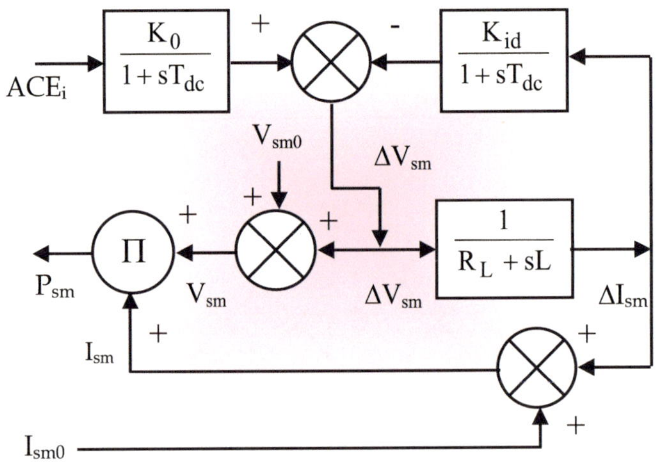

Fig. 10. Superconducting magnetic energy storage unit control system

7. Simulation results

To demonstrate the usefulness of the proposed controller, computer simulations were performed using the MATLAB environment under different operating conditions. The system performances with gain scheduled SMES and fixed gain SMES are shown in Fig. 11 through Fig. 14. Two case studies are conducted as follows:

Case I: a step load increase (ΔP_{L2}=0.01 pu) is considered in area2 only.

It is seen from Fig. 11 that, the tie line power deviation are more reduced with the proposed gain scheduled controller than the fixed gain one including SMES, and the deviations are positive in Case I. Thus sensing the input signal ACE_i in both the control areas SMES provide sufficient compensation as shown in Fig. 12, where in area1 SMES is charging/discharging energy and area2 SMES is discharging/charging energy to keep the frequency deviations in both areas minimum. From Fig. 12 it is seen that, fuzzy gain scheduled integral controller of the loaded area determines the integral gain, K_I, to a scheduled value to resotore the frequency to its nominal value, and fuzzy gain scheduled integral controller of the unloaded area reamains unscheduled and selects the critical value

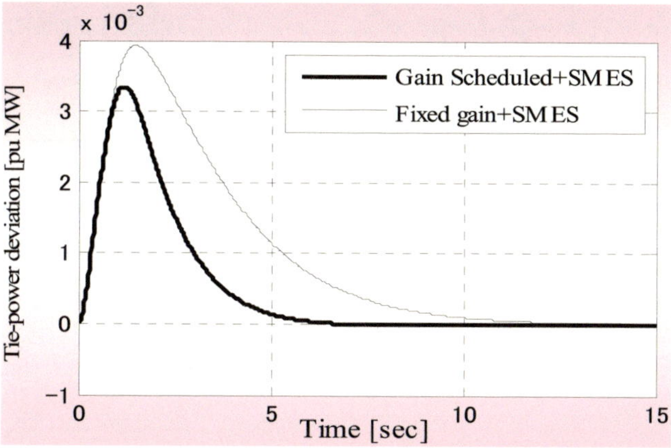

Fig. 11. Performances of tie power deviation for a step load increase ΔP_{L2}=0.01 pu in area2 only

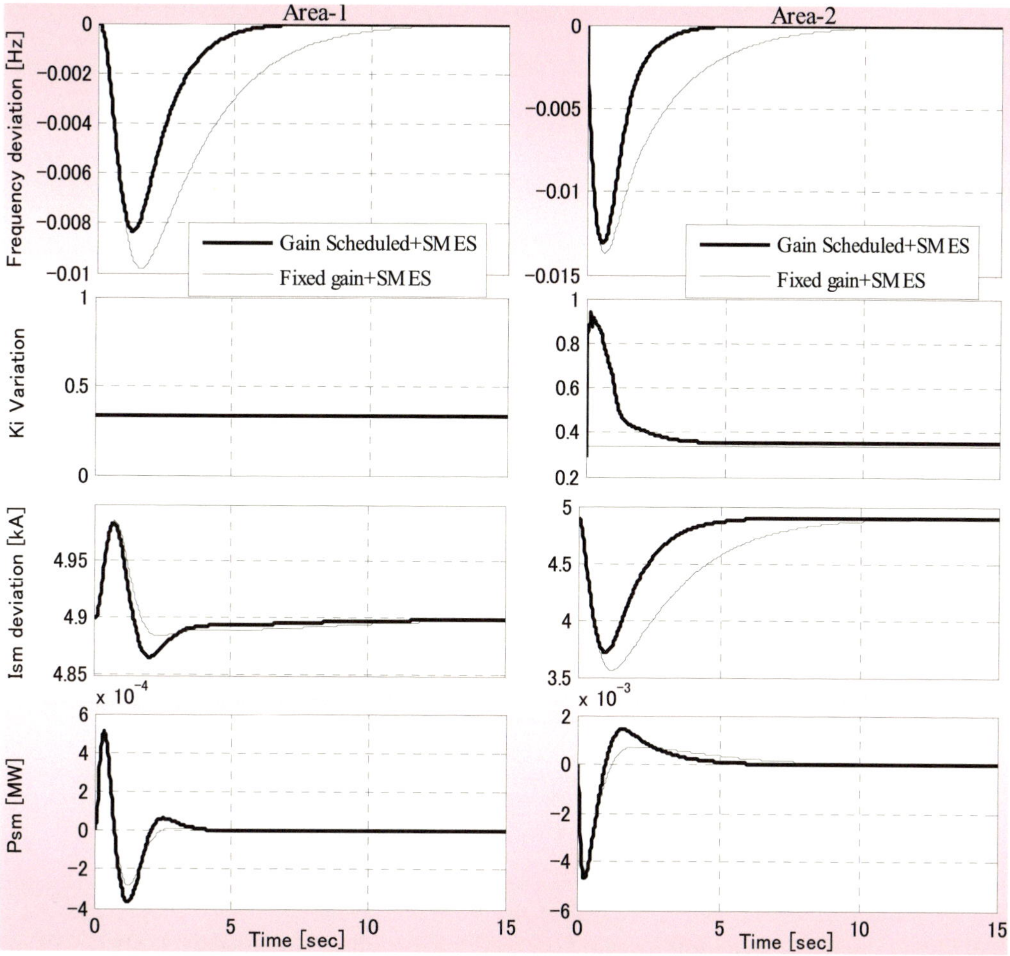

Fig. 12. System performances for a step load increase ΔP_{L2}=0.01 pu in area2 only

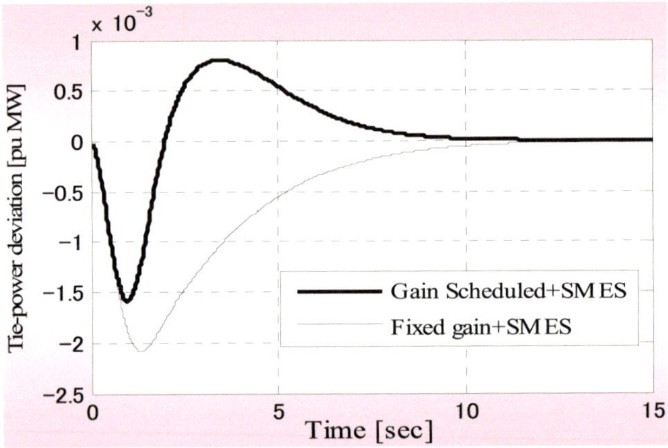

Fig. 13. performances of tie power deviation for a step load increase ΔP_{L1}=0.015 pu in area1 & ΔP_{L2}= 0.01 pu in area2

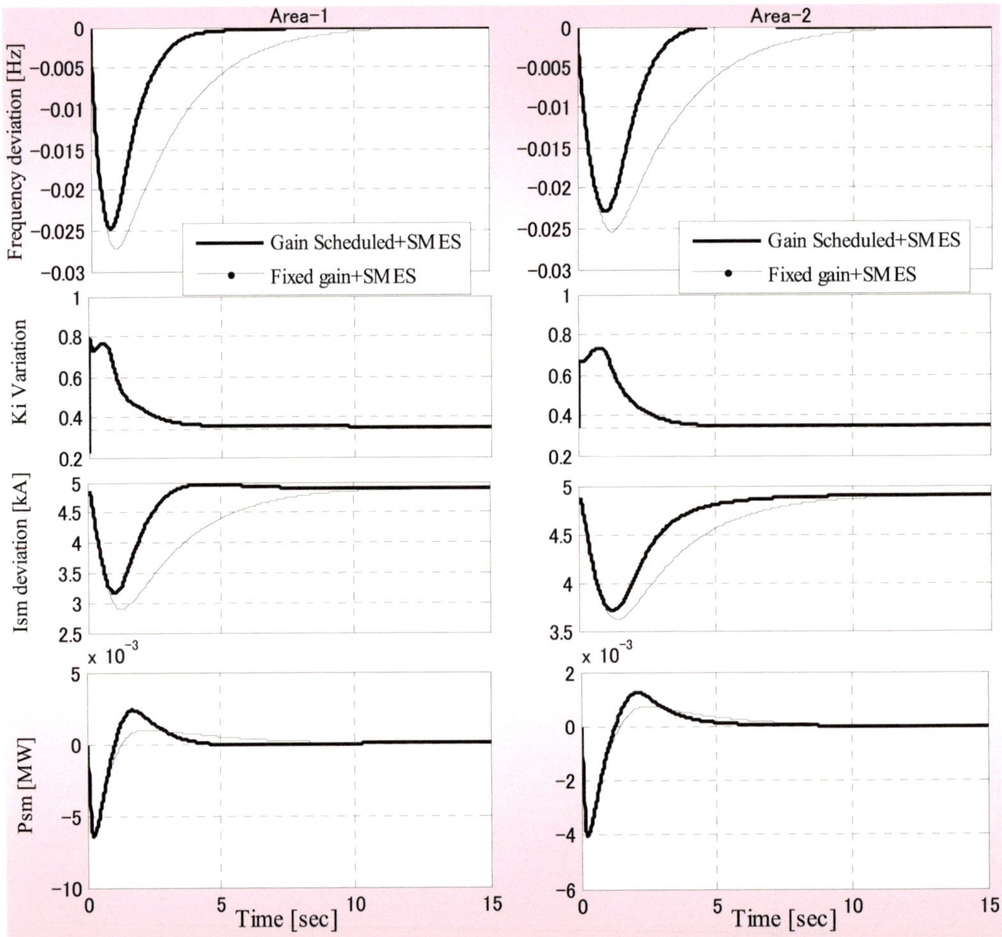

Fig. 14. System performances for a step load increase ΔP_{L1}=0.015 pu in area1 & ΔP_{L2}= 0.01 pu in area2

as its integral gain. In addition, it is seen that, the damping of the system frequency is not satisfactory in the case with the fixed gain controller including SMES, but the proposed gain scheduled supplementary controller including SMES significantly improves the system performances.

***Case II*:** different step load increase is applied to each area.

In this case, as each area is loaded by the different load increase, each area adjusts their own load. Fig. 13 shows the tie power deviation but the magnitude is small. So the SMES controller in both areas dominated on Δf_i. As ΔP_{L1}=0.015 pu & ΔP_{L2}=0.01 pu, it is seen from Fig. 14 that SMES in area1 provided more compensation than that in area2. The inductor current deviation (ΔI_{sm}) is also reduced significantly and return back to the rated value quickly with the proposed control system. Finally, it is seen from Fig. 14 that fuzzy gain scheduled integral controller of both the loaded areas determine the integral gain K_{Ii} to a scheduled value to resotore the frequency to its nominal value. Due to this, the damping of the system frequency is also improved with the proposed FGSPI controller including SMES.

8. Chapter conclusions

The chapter discussed about the simulation studies that have been carried out on a two-area power system to investigate the impact of the proposed intelligently controlled SMES on the improvement of power system dynamic performances. The results clearly show that the scheme is very powerful in reducing the frequency and tie-power deviations under a variety of load perturbations. On-line adaptation of supplementary controller gain associated with SMES makes the proposed intelligent controllers more effective and are expected to perform optimally under different operating conditions.

9. References

Benjamin, NN. & Chan, WC. (1978). Multilevel Load-frequency Control of Inter-Connected Power Systems, *IEE Proceedings, Generation, Transmission and Distribution*,Vol. No.125, pp.521-526.

Nanda, J. & Kavi, BL. (1988). Automatic Generation Control of Interconnected Power System, *IEE Proceedings, Generation, Transmission and Distribution*, Vol. 125, No. 5, pp.385-390.

Das, D.; Nanda, J.; Kothari, ML. & Kothari, DP. (1990). Automatic Generation Control of Hydrothermal System with New Area Control Error Considering Generation Rate Constraint, *Electrical Machines and Power System*, Vol. 18, pp.461-471.

Mufti, M. U.; Ahmad Lone, S.; Sheikh, J. I. & Imran, M. (2007). Improved Load Frequency Control with Superconducting Magnetic Energy Storage in Interconnected Power System, *IEEJ Transactions on Power and Energy*, Vol. 2, pp. 387-397.

Nanda, J.; Mangla, A & Suri, S. (2006). Some New Findings on Automatic Generation Control of an Interconnected Hydrothermal System with Conventional Controllers, *IEEE Transactions on Energy Conversion*, Vol. 21, No. 1, pp. 187-194, (March, 2006).

Oysal, Y.; Yilmaz, A.S. & Koklukaya, E. (2004). Dynamic Fuzzy Networks Based Load Frequency Controller Design in Electrical Power Systems, *G.U. Journal of Science*, Vol. 17, No. 3, pp. 101-114

Benjamin, NN. & Chan WC. (1982). Variable Structure Control of Electric Power Generation. *IEEE Transactions on Power Apparatus and System*, Vol. 101, No. 2, pp.376-380.

Sivaramaksishana, AY.; Hariharan, MV. & Srisailam, MC. (1984). Design of Variable Structure Load-Frequency Controller Using Pole Assignment Techniques, *International Journal of Control,* Vol. 40, No. 3, pp.437–498.

Tripathy, SC, & Juengst, KP. (1997). Sampled Data Automatic Generation Control with Superconducting Magnetic Energy Storage, *IEEE Transactions on Energy Conversion* Vol. 12, No. 2, pp.187–192.

Shayeghi, H. & Shayanfar, H.A. (2004). Autometic Generation Control of Interconnected Power System Using ANN Technique Based on μ-Synthesis, *Journal of Electrical Engineering,* Vol. 55, No. 11-12, pp. 306-313.

Sheikh, M.R.I.; Muyeen, S.M.; Takahashi, R.; Murata, T. & Tamura, J. (2008). Improvement of Load Frequency Control with Fuzzy Gain Scheduled Superconducting Magnetic Energy Storage Unit, *International Conference of Electrical Machine (ICEM, 08),* (06-09 September, 2008), Vilamura, Portugal.

Demiroren, A. (2002). Application of a Self-Tuning to Automatic Generation Control in Power System Including SMES Units, *European Transactions on Electrical Power,* Vol. 12, No. 2, pp. 101-109, (March/April 2002).

IEEE Task Force on Benchmark Models for Digital Simulation of FACTS and Custom–Power Controllers, T&D Committee, (2006). Detailed Modeling of Superconducting Magnetic Energy Storage (SMES) System, *IEEE Transactions on Power Delivery,* Vol. 21, No. 2, pp. 699-710, (April 2006).

Ali, M. H.; Murata, T. & Tamura, J. (2008). Transient Stability Enhancement by Fuzzy Logic-Controlled SMES Considering Coordination with Optimal Reclosing of Circuit Breakers, *IEEE Transactions on Power Systems,* Vol. 23, No. 2, pp. 631-640, (May 2008).

http://www.doc.ic.ac.uk/~matti/ise 2grp/energystorage_report/node8.html

http://en.wikipedia.org/wiki/Superconducting_magnetic_energy_storage

Demiroren, A. & Yesil, E. (2004). Automatic Generation Control with Fuzzy Logic Controllers in the Power System Including SMES Units, *International Journal of Electrical Power & Energy Systems,* Vol. 26, pp. 291-305.

Abraham, R.J.; Das, D. & Patra, A. (2008). AGC Study of a Hydrothermal System with SMES and TCPS, *European Transactions on Electrical Power,* DOI: 10.1002/etep.235

Wu, C. J. & Lee, Y. S. (1991). Application of Superconducting Magnetic Energy Storage to Improve the Damping of Synchronous Generator, *IEEE Transactions on Energy Conversion,* Vol. 6, No. 4, pp. 573-578, (December 1991).

Banerjee, S.; Chatterjee, J. K. & Tripathy, S. C. (1990). Application of Magnetic Energy Storage Unit as Load Frequency Stabilizer, *IEEE Transactions on Energy Conversion,* Vol. 5, No. 1, pp. 46-51, (March 1990).

M.R.I. Sheikh was born in Sirajgonj, Bangladesh on October 31, 1967. He received his B.Sc. Eng. and M.Sc. Eng. Degree from Rajshahi University of Engineering & Technology (RUET), Bangladesh, in 1992 and 2003 respectively, all in Electrical and Electronic Engineering. He is currently an Associate Professor in the Electrical and Electronic Engineering Department, RUET. Presently he is working towards his Ph.D Degree at the Kitami Institute of Technology, Hokkaido, Kitami, Japan. His research interests are, Power System Stability Enhancement Including Wind Generator by Using SMES, FACTs devices and Load Frequency Control of multi-area power system.

Mr. Sheikh is the member of the IEB and the BCS of Bangladesh.

6

Influence of Streamer-to-Glow Transition on NO Removal by Inductive Energy Storage Pulse Generator

Koichi Takaki
Iwate University
Japan

1. Introduction

Huge amounts of air pollutants like carbon monoxide, unburned hydrocarbons, nitrogen oxides (NOx), and particulate matter have been released into the atmosphere by various sources such as coal, oil, and natural gas-burning electric power generating plants, motor vehicles, diesel engine exhaust, paper mills, metal and chemical production plants, etc., over the last several decades. These pollutants are the main cause of acid rain, urban smog, and respiratory organ disease (Chang, 2001). For pollutants emitted from motor vehicle, the exhaust of gasoline engines is cleaned effectively with the three-way catalyst. However, for diesel and lean burn engines, the three-way-catalyst does not work because the high oxygen content in the exhaust gases prevents the reduction of nitrogen oxide (NO) (Clements et al., 1989).

Dry NOx removal technology is one of the conventional processes which may provide a potential solution for such problems (Eliasson and Kogelschatz, 1991). A non-thermal plasma process using a pulse streamer corona discharge is particularly attractive for this purpose (Namihira et al., 2000). During the past decade, numerous studies on this process have been conducted using a diesel engine exhaust gas and/or a simulated gas (Hackam & Akiyama, 2000). Although encouraging results have been obtained from the experiments, it is urgent to design a whole removal system compact enough for vehicle application.

Two methods for storing energy are employed in high-power pulse generators: capacitive and inductive storages. When the energy is stored in capacitors, the energy is transferred to a load through closing devices, e.g., high-current nanosecond switches. If the energy is stored in an inductive circuit with current, opening switch is used to transfer energy to a load (Rukin, 1999). For short-pulsed high voltage generation with high impedance load, inductive energy storage (IES) system is more adequate than capacitive energy storage system, if appropriate opening switches are available (Jiang et al., 2007).

High-voltage nanosecond pulse generators, in which high-voltage semiconductor diodes are employed for interrupting currents stored as inductive energy, have been developed (Rukin, 1999). The generators using the high-voltage diodes as semiconductor opening switch (SOS) have an all-solid-state switching system and therefore, combine high pulse repetition rate, stability of the output parameters and long lifetime (Grekhov & Mesyats, 2002). SOS pulse generators operating at various institutions demonstrated their high reliability during

applied research work connected with the pumping of gas lasers (Baksht et al., 2002), ionization of air with a corona discharge (Yalandin, et al., 2002, Cathey, et al., 2007), generation of radical species with a atmospheric pressure glow discharge (Takaki, et al., 2005), and generation of high-power microwave (Bushlyakov et al., 2006).
The streamer discharges driven by a pulsed power generator can dissociate oxygen molecules to atomic oxygen radicals with high-energy efficiency because of low-conductive current loss (Fukawa et al., 2008). The IES pulsed power generator using SOS diodes is particularly attractive for this purpose because the whole system can be compact, lightweight and driven at high repetition rate. However, a discharge produced by the IES pulsed power generator transients from streamer to glow when the energy stored in the capacitor still remains after the energy transfer from a capacitor to an inductor at opening the SOS diodes (Grekhov & Mesyats, 2002). As the results, the energy efficiency for gas treatment using non-thermal plasma is affected by the streamer-to-glow transition (Takaki et al., 2007). In here, NO removal using a co-axial type non-thermal plasma reactor driven by an IES pulsed power generator is described. The influence of streamer-to-glow transition on NO removal in the non-thermal plasma reactor is also described.

2. Experimental setup

Figure 1(a) shows the schematics of the experimental circuit. The IES pulsed power generator consists of a primary energy storage capacitor *C*, a closing switch SW, a secondary energy storage inductor *L*, and an opening switch. The circuit current flows to the LC circuit governed by the following equation after closing the switch SW (Robiscoe et al., 1998):

$$i = \frac{V_0}{\omega_0 L} e^{-\frac{R}{2L}t} \sin\omega_0 t \,, \tag{1}$$

$$\omega_0 = \sqrt{\frac{1}{LC} - \left(\frac{R}{2L}\right)^2} \,, \tag{2}$$

where t is the time from the activation of the closing switch, V_0 is the charged voltage, L is the inductance of the energy storage inductor, C is the capacitance of the primary energy storage capacitor, and R is the circuit resistance ($R < 4\, L / C$). When SOS diodes are used as an opening switch as shown in Figure 1(a), the circuit current flows through the SOS diodes as a forward-pumping current during a half period $T_F \approx \pi\sqrt{LC}$ of LC oscillation (Yalandin et al., 2000). After the current direction reverses with LC oscillation, the reverse current is injected into the SOS during the period T_R. After the injection phase T_R, the circuit current is interrupted by a short duration T_O. With the current interrupted by the SOS, a high-voltage pulse is produced as follows:

$$V_{out} = V_0 - \frac{1}{C}\int i dt - L\frac{di}{dt} - Ri \approx -L\frac{di}{dt} \,, \tag{3}$$

as shown in Fig. 1(b). This pulse voltage can be applied to a load as a short nanosecond pulse (Takaki et al., 2005, Rukin, 1999, Yankelevich & Pokryvailo, 2002).

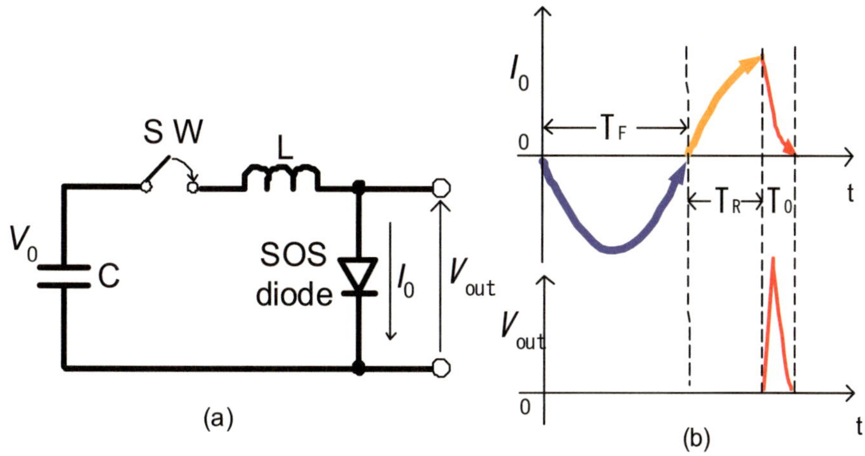

Fig. 1. Schematic of an inductive energy storage pulse power generator with semiconductor opening switch: (a) equivalent circuit; (b) circuit current and output voltage.

Fast reverse recovery diodes VMI K100UF (Voltage Multipliers, Inc., 10 kV maximum voltage, 100 A maximum current, T_R=100 ns) were employed as SOS diodes, i.e., a semiconductor opening switch. K100UF diodes were connected in 5 series and 4 parallel to decrease the circuit inductance and to increase capable forward pumping current and reversed voltage (400 A maximum current, 50 kV maximum voltage). The capacitance of the primary energy storage capacitor *C* and the inductance of the secondary energy storage inductor *L* were changed in range from 0.12 to 4.2 nF and from 4.8 to 18.5 μF, respectively, as shown in **Table 1**. The charging voltages of the capacitor *C* were -20 kV for conditions 1-3 and -12 kV for condition 4. The pulse repetition rate was changed to control the input energy in the reactor. The current and voltage were measured with Pearson 2878 current transformers (0.1 V/A sensitivity, 400 A maximum current, 5 ns rise time) and Tektronix high-voltage P6015A probe (40 kV peak voltage, 4 ns rise time), respectively. The signals stored in a Tektronix TDS3054B digitizing oscilloscope (500 MHz band width, 5 GS/s sampling rate) was transmitted to a computer through a LAN cable for calculating the energy consumed in reactor.

Condition	#1	#2	#3	#4
C [nF]	0.12	0.23	0.48	4.2
L [μH]	18.5	10.5	4.8	12.6
T_F [ns]	203	205	224	876
T_R [ns]	67	68	69	102

Table 1. Forward and reversed pumping time of SOS diodes for various circuit parameters.

Figure 2 shows a schematic of the experimental set-up using the pulse streamer discharge reactor. The simulated gas was diluted NO with nitrogen and oxygen mixed with ratio of 9:1. The co-axial plasma reactor consists of a 1mmϕ tungsten wire and a copper cylinder

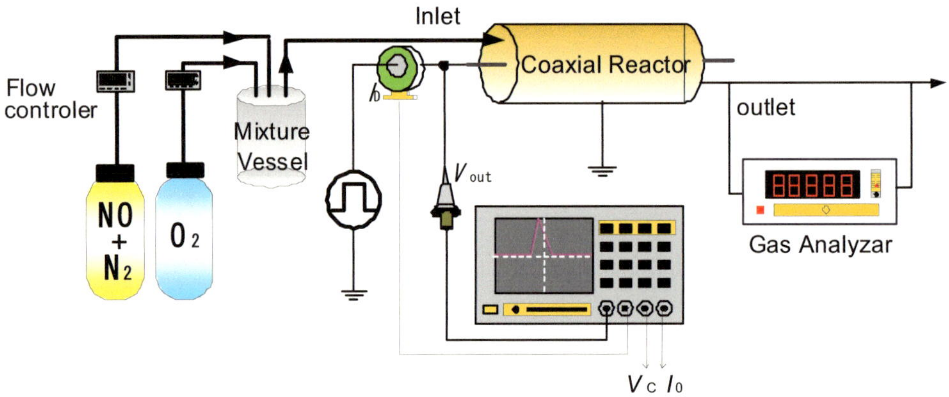

Fig. 2. Experimental setup for NO removal from simulated exhaust gas using a corona discharge reactor driven by an IES pulse power generator.

with an inner diameter of 20 mm. The length of the reactor is 300 mm, which corresponds to 94 cc in volume. The initial concentration of NO gas was controlled to 200 ppm using a mass flow controller. The flow rate of the simulated gas was changed from 2 to 4 L/min. The NO and NO_2 gases were analyzed by Best Sokki BCL-511 gas analyzer.

3. Results and discussion

3.1 Pulse power circuit behavior

Figure 3 shows typical waveforms of circuit current I_0, capacitor voltage V_C, and output voltage V_{out} without connecting to the load at C=0.48 nF and L=4.8 μH. The charging voltage V_0 of the capacitor C is -10 kV. "Time = 0" means the time after closing the gap switch SW. The circuit current I_0 starts to flow after closing the gap switch SW with LC oscillation. The diode forward-pumping period T_F, i.e., a half period of LC oscillation, is 210 ns, and the peak of the forward-pumping current is 67 A. After the current direction reverses with LC oscillation, the reverse current is injected into the SOS during a 70 ns period T_R. After the 70 ns injection phase T_R, the circuit current is interrupted for 25 ns duration T_O. During this phase, the 50 A reversed current is interrupted for 25 ns. The output voltage increases rapidly and has a maximum voltage of 22 kV, which corresponds to 2.2 of an amplification factor, i.e. corresponds to the ratio of the maximum output voltage to the charging voltage V_0. The pulse width of the output voltage is 25 ns in FWHM (full-width at half-maximum). The total inductance of the circuit is the summation of the secondary energy storage inductor and a circuit loop inductance. The total circuit inductance can be roughly estimated using the equation $T_F \approx \pi\sqrt{LC}$ and is calculated to be 9.6 μH using a 210 ns half period of LC oscillation. This inductance and rapid current interruption produces a high voltage pulse expressed by Equation (3).

Table 1 shows the periods of a forward-pumping current T_F and the periods of a reverse current of the diodes T_R for various circuit conditions. The period T_F has approximately same values at around 200 ns (203, 205, and 224 ns for conditions 1, 2, and 3, respectively) for all three different conditions. The period T_F can be predicted as a half cycle of LC oscillation and was calculated to be 146, 154, and 150 ns using values of C and L of the conditions 1, 2, and 3, respectively. The measured period T_F shows a larger value than those

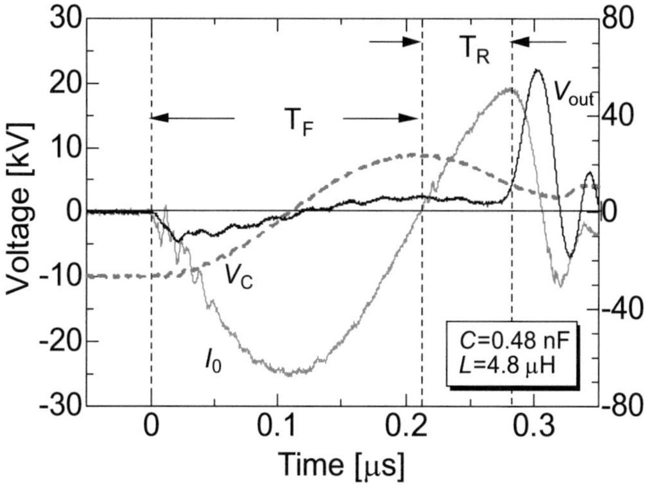

Fig. 3. Typical waveforms of circuit current I_0, capacitor voltage V_C, and output voltage V_{out} without connection to the load at C=0.48 nF, L=4.8 μH and V_0=-10 kV.

calculated by the result of a stray inductance of the circuit. The reverse current period T_R has approximately the same values for different circuit conditions in the same forward current period. However, the value of T_R is around 70 ns which is 70% of the rated period of 100 ns. The values of L and C in condition 4 were chosen to increase the forward-current period T_F as $\pi\sqrt{LC} = 723$ ns. The obtained period of T_R is 102 ns, which agreed well with the rated period of 100 ns.

Figure 4 shows the typical time-dependency of the discharge current I_{load}, the circuit current I_0, the reactor voltage V_{out} and the voltage of the primary energy storage capacitor V_C with connection of the pulsed power generator to the reactor. The circuit condition is chosen as C=0.68 nF and L=1.4 μH. The charging voltage V_0 of the capacitor is set to be -20 kV. The discharge current I_{load} can be divided by two parts; displacement current at the early part of the current and a discharge current as the sharp peak. The peak value of the discharge current is around 100 A. The voltage of the capacitor V_C is almost zero when the pulse voltage is produced and applied to the reactor. This result indicates the energy stored in the capacitor is almost released through LC oscillation.

Figure 4 also shows time-dependency of energy stored in the secondary energy storage inductor E_L, energy stored in the primary energy storage capacitor E_C, energy loss in the SOS diodes E_{SOS} and energy consumed in the reactor E_{load}. The energy stored in the capacitor is transferred to the inductor in the period of first quarter cycle of the LC oscillation. After that, the energy stored in the inductor is transferred back to the capacitor in next quarter cycle. The energy of 16 mJ is consumed in the SOS in the period of LC half cycle because of the resistive component of the SOS diodes. The total energy transfer from the primary energy storage capacitor to the reactor is around 20% under the circuit condition. The energy transfer efficiency changes by changing circuit parameter as reported in the reference (Takaki et al., 2007). The energy transfer efficiency increases to 40% by decreasing circuit current from 200 to 50 A in peak value because the energy loss in the SOS decreases with decreasing current through the SOS diodes.

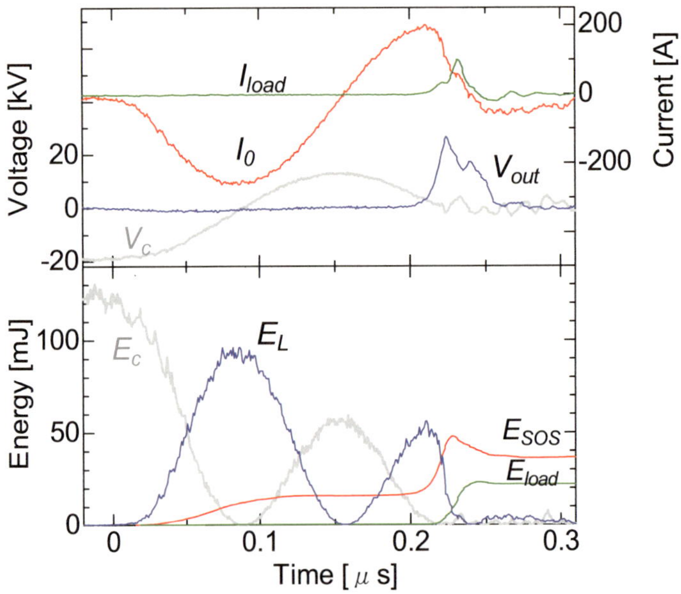

Fig. 4. Typical time-dependency of the discharge current I_{load}, the circuit current I_0, the reactor voltage V_{out} and energy consumed in SOS, E_{SOS} and in the reactor E_{load} at C=0.68 nF, L=1.4 μH and V_{C0}=-20 kV. E_C and E_L mean energies stored in the capacitor and the inductor, respectively.

Figure 5 shows the time dependence of the discharge current, the voltage between the electrodes, and the energy consumed for different circuit conditions. The energy consumed was calculated based on a multiplied value of the discharge current and the voltage (Takaki et al., 2000). The peak voltages between the electrodes were 19, 25, 27 kV for conditions 1, 2, and 3, respectively. The peak current also increased from 32 to 54 A with an increase in the capacitance C from 0.12 to 0.48 pF. The energies consumed in the discharge are 9, 13, 19 mJ for the circuit conditions 1, 2, and 3, respectively. These values correspond to 39, 28, and 20% of the primary energy obtained by $0.5CV_0^2$, where V_0 is the charging voltage of the storage capacitor C.

Figure 6 shows the waveforms of the discharge current, the voltage between the electrodes, and the energy consumed for circuit condition 4. The time transient of circuit condition 2 is also shown in Fig. 6 for the comparison of different circuit conditions. The peak voltage and current obtained were 25 kV and 62 A, respectively. The waveforms of the discharge current and voltage have a long decay time compared to the cases of the conditions 1 to 3 as shown in Fig. 5. The time constant of the current decay is approximately 300 ns. The time constant can be expressed as R_d x C, where R_d represents an equivalent resistance of the discharge. The equivalent resistance R_d is 70 Ω using the capacitance 4.2 nF as the value of C. This longer discharge current decay may indicate that a glow discharge is generated with the remaining electrical charges in the storage capacitor (K. Takaki et al., 2005). The voltage between the electrodes after the diodes interrupt the circuit current still remains around 11.5 kV. This voltage corresponds to 48 μC of the remaining charge in a storage capacitor. The total energy consumed is around 125 mJ which consists of 14 mJ in energy consumption by the pulse streamer discharge (at 73 ns from applying voltage, just prior to a quasi-stable phase) and 111 mJ by the quasi-stable glow discharge (=125 - 14 mJ). The 14 mJ energy consumed by the streamer discharge is similar to the values obtained under the circuit conditions 1to 3 as shown in Fig. 5.

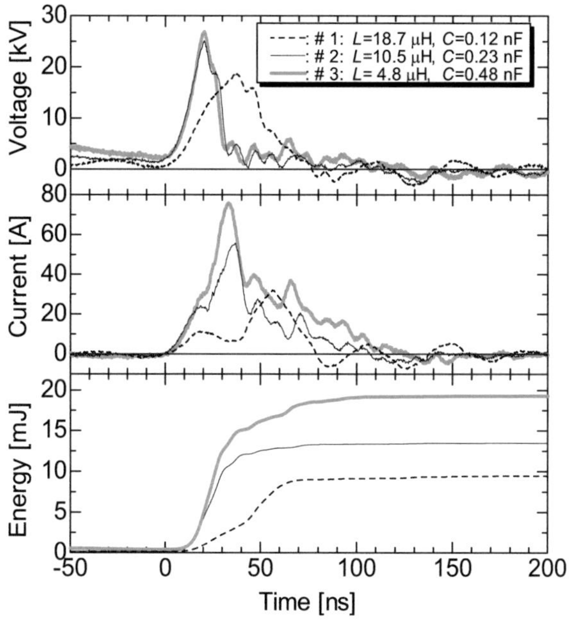

Fig. 5. Waveforms of applied voltage, discharge current, and energy consumed in the reactor for three different circuit conditions.

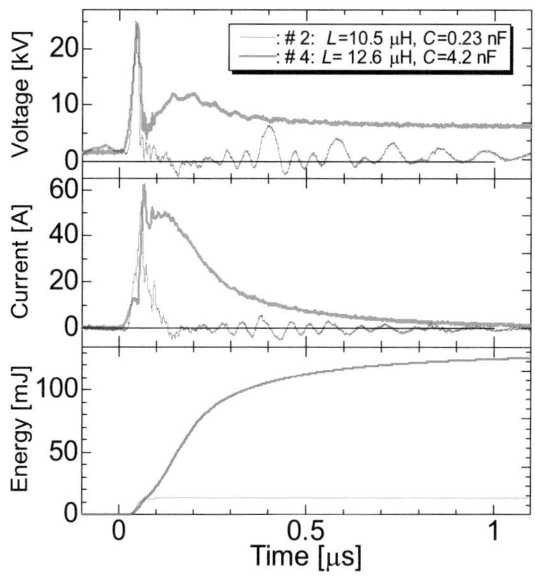

Fig. 6. Waveforms of applied voltage, discharge current, and energy consumed in the reactor at 4.2 nF capacitance.

3.2 NO removal efficiency

Figure 7 shows the NO removal efficiency as a function of the pulse repetition rate for two different gas flow rates and various circuit conditions. The charging voltages of the primary storage capacitor are maintained at around 25 kV of the output voltage and are set to be -20 kV for conditions 1 to 3 and -12 kV for circuit condition 4. The pulse repetition rate was controlled by the amplitude of charging current to the primary capacitor. NO removal increases with an increase in the pulse repetition rate for all cases. NO removal also increases with decreasing gas flow rate from 4 to 2 L/min. In all cases, NO_x (=NO+NO_2) removal was found to be lower than 10% by additional measurement using the NO_x analyzer (Best Sokki BCL-511). This indicates that the NO removal is mainly due to oxidization of NO to NO_2. The oxidation reactions can be written as follows:

$$NO + O_3 \rightarrow NO_2 + O_2, \quad (4)$$

$$NO + O + M \rightarrow NO_2 + M, \quad (5)$$

where M indicates ambient gas. The rate coefficient of reaction (4) k_{NO2} is reported to be 1.7x10^{-14} cm^3s^{-1} at 27 centigrade degree (Hill et al., 1988, Lowke and Morrow, 1995). The oxidation rate is expressed as k_{NO2} [NO][O_3], where [NO] and [O_3] indicate concentrations of NO and ozone, O_3, respectively. The ozone is generated with the reaction between an oxygen molecule and an atom of oxygen which is mainly generated via energetic electron collision with the oxygen molecule. The reaction is expressed as follows;

$$e + O_2 \rightarrow 2O + e, \quad (6)$$

$$O + O_2 + M \rightarrow O_3 + M. \quad (7)$$

where M indicates ambient gas. The rate coefficient of the reaction (7) k_{O3} is 6.3x10^{-34} $cm^{-6}s^{-1}$ at 300 K (Vikharev et al., 1993, Chang et al., 1991).

Some researchers employed input energy density (specific energy density; SED) to evaluate scale-up of NO removal yield (Hackam & Akiyama, 2000, Penetrante et al., 1995).

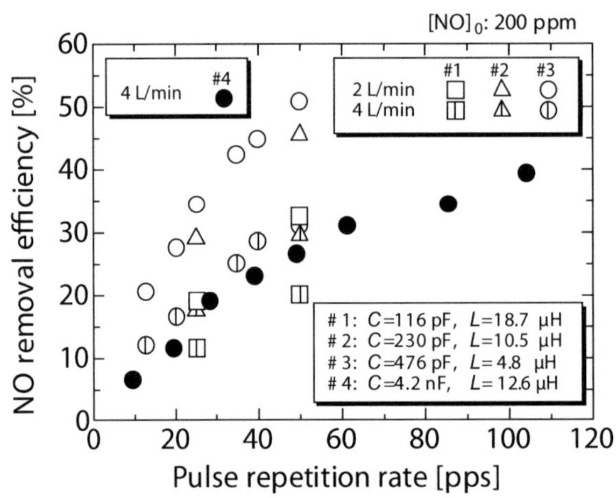

Fig. 7. Dependence of NO removal efficiency on pulse repetition rate for various gas flow rates and circuit parameters.

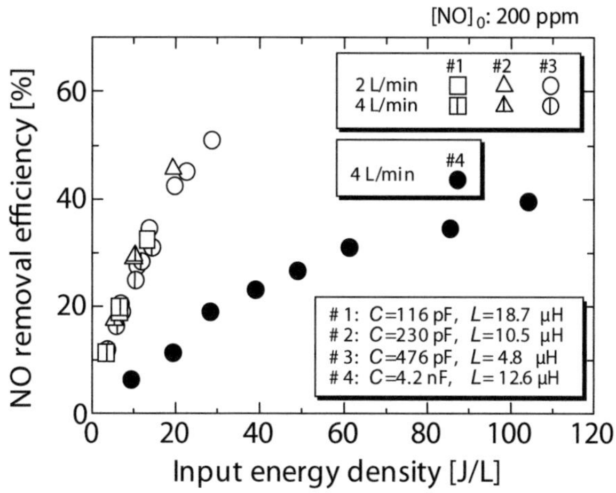

Fig. 8. Dependence of NO removal efficiency on the energy density for various circuit conditions.

Figure 8 shows the NO removal efficiency as a function of input energy density in the reactor for various circuit conditions. The input energy density is determined by the following equation,

$$SED = \frac{Input \ \ electric \ \ power}{Gas \ \ flow \ \ rate} = \frac{60 \cdot U \cdot f}{q} \ \ [\text{J/L}], \tag{8}$$

where U means energy consumed in the reactor per one pulse of applied voltage obtained, as shown in Figs. 5 and 6. f and q are the pulse repetition rate [Hz] and the gas flow rate [L/min.], respectively. 1 [J/L] corresponds to 3.6 [Wh/m^3]. NO removal increases with input energy density. NO removal values of conditions 1 to 3 have similar values, whereas the NO removal value of condition 4 is much lower than those of other conditions. This result indicates that the primary energy storage capacitor C and the secondary energy storage inductor L must be set at $T_F/2 \approx T_R$ $(\approx 100ns)$ for higher NO removal with lower energy input. When the values of C and L are set at $T_F/2 << T_R$, the peak voltage of the pulsed power generator has a lower value owing to the lower di/dt value (Takaki & Akiyama, 1992). When the C and L values are set at $T_F/2 >> T_R$, where the quasi-stable discharge (glow mode) is generated after the pulse streamer discharge. Another reason of the difference between NO removal for conditions 1 to 3 and condition 4 may be due to difference of the charging voltage for the primary energy storage capacitor. The voltages were -12 kV for conditions 1 to 3 and -20 kV for condition 4. However, this difference is considered to only slightly affect the NO removal during the streamer mode because the peak voltage and the discharge current have slight differences between condition 2 and condition 4 as shown in Fig. 6.

Figure 9 shows the energy efficiency of NO removal as a function of the NO removal rate for various circuit conditions. The energy efficiency of NO removal is determined by the following equation,

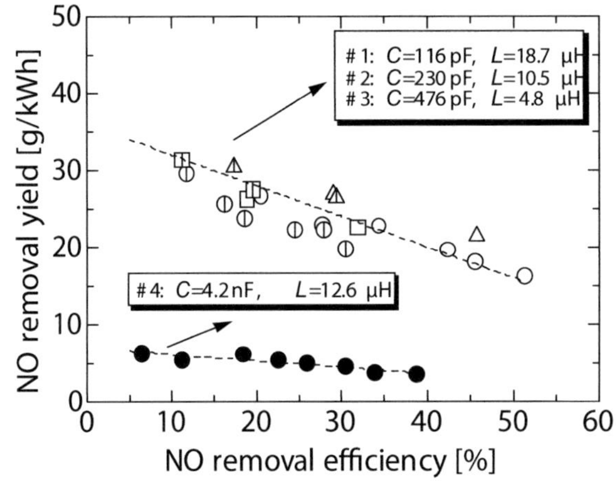

Fig. 9. Dependence of NO removal energy efficiency on NO removal for various circuit conditions.

$$\eta = \frac{60 \cdot q \cdot \Delta[NO] \cdot M}{N \cdot P} \times 10^3 \text{ [g/kWh]}, \quad (9)$$

where Δ[NO] is the NO removal [ppm], M is the NO molecular weight (=30 g), N is the molar volume (=22.4 L), and P is the power consumed in the reactor (= U x f [W]). The energy efficiencies of conditions 1 to 3 have similar values around 25 g/kWh at 30% of NO removal. However, the energy efficiency of condition 4 is much lower than that of other circuit conditions at 5 g/kWh. The NO removal energy efficiency of 25 g/kWh is similar order of magnitude with a pulse corona or pulsed streamer plasma reactor (Hackam & Akiyama, 2000, Dinelli et al., 1990, Amirov et al., 1993). However, well-optimized pulse streamer plasma reactors show higher energy efficiency than present system (Namihira et al., 2000). For example, Yankelevich et al. (2006) reported NO removal energy efficiency of 30 g/kWh for 30% of NO removal at 430 ppm initial NO concentration in diesel engine exhaust (around 12.5% O_2 in the gas mixture). More optimization of present system is required to improve its efficiency.

For condition 4, the current waveform of the pulse driven glow discharge as shown in Fig. 6 can be divided into three parts: displacement current, streamer discharge current, and glow discharge current (Takaki et al., 2000). In the present experimental condition, the displacement current and the streamer discharge current appear around 40 ns and 65 ns after opening the switch diodes, as shown in Fig. 6. The glow discharge appears around 80 ns after opening the switch and is sustained about 1 μs. The glow discharge voltage after 200 ns from opening the switch is 10 kV. The value of 10 kV corresponds to 37 Td in the reduced electric field E/N (E: electric field, N: gas density) of a positive column area using the values 1 cm gap length and 285 V cathode sheath drop (Takaki et al., 2000). The electron temperature T_e can be estimated from swarm parameters of electrons in nitrogen according to Einstein's well-known equation,

$$k_B T_e / e \approx D_e / \mu_e \quad (10)$$

where μ_e, k_B and D_e are drift mobility, Boltzmann constant (1.38×10^{-23} JK^{-1}), and the diffusion constant, which can be expressed as a function of *E/N* (Nakamura, 1987). As a result, the electron temperature is calculated to be 1.3 eV, which is much lower than the typical 2 to 5 eV values of the streamer discharge. The reaction (6) requires higher electron energy larger than 6.1 eV. The low electron temperature of the glow phase compared to the streamer discharge is one of the reasons for lower energy yield for NO removal in condition 4.

3.3 Influence of oxygen concentration on energy transfer

Figure 10 shows time-dependence of a discharge current and a voltage between the electrodes for various oxygen concentrations at circuit condition 3; *C*=0.48 nF and *L*=4.8 μH shown in Table 1. The charging voltage of the primary capacitor is -20 kV. The flow of the gas mixture into the reactor is 4 L/min. The rise time of the voltage between the electrodes has almost the same value for all oxygen concentrations. As a result, the displacement current which is the first part of the current waveform has the same value at all oxygen concentrations. The discharge current appears around 35 ns following by the displacement current. The discharge current is quenched earlier by increasing oxygen concentrations. However, the discharge current in all cases is quenched in short time, lower than 100 ns.

Figure 11 shows time-dependence of a discharge current and a voltage between the electrodes for various oxygen concentrations at the circuit condition 4, *C*=4.2 nF and *L*=12.6 μH. The charging voltage and the flow rate of the gas mixture are -10 kV and 4 L/min, respectively. In this circuit condition, the displacement current also has the same value at all oxygen concentrations. The discharge current appears at around 80 ns followed by the displacement current. When the oxygen concentration is lower than 25%, the waveforms of the discharge current and voltage have long decay time compared with the circuit condition 3 as shown in Figure 10. The decay time decreases with increasing oxygen concentrations.

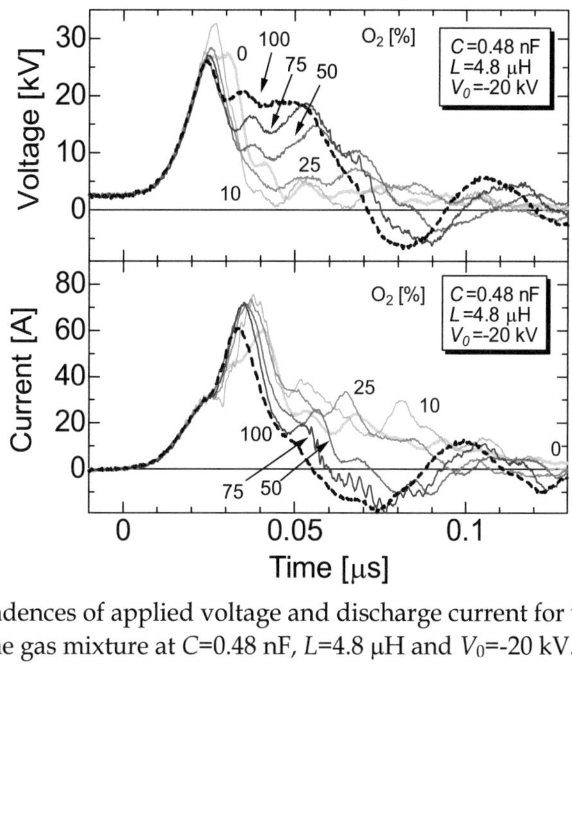

Fig. 10. Time-dependences of applied voltage and discharge current for various oxygen concentrations in the gas mixture at C=0.48 nF, *L*=4.8 μH and V_0=-20 kV.

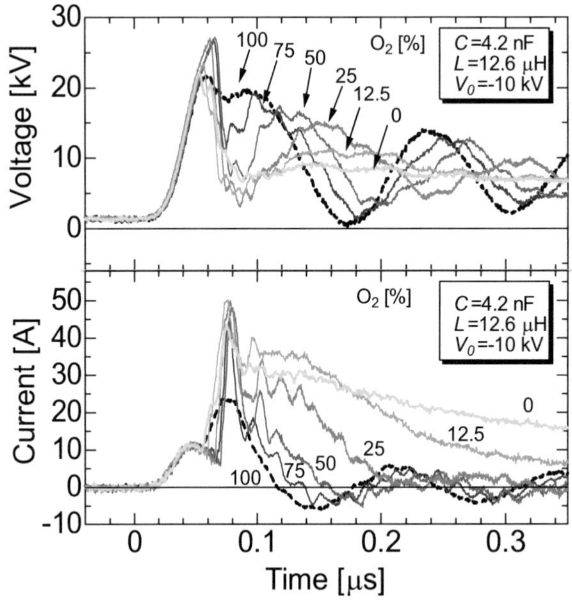

Fig. 11. Time-dependences of applied voltage and discharge current for various oxygen concentrations in the gas mixture at C=4.2 nF, L=12.6 μH and V_0=-10 kV.

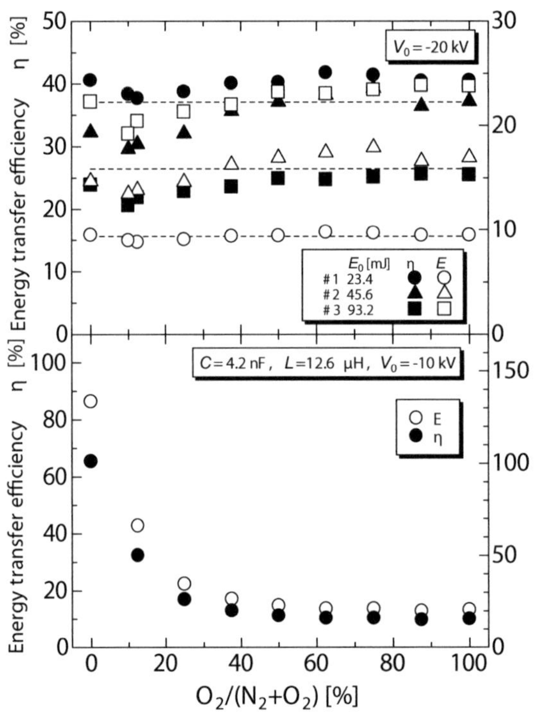

Fig. 12. Energy consumed in the reactor and the energy transfer efficiency from the primary capacitor to the reactor as a function of oxygen concentration for various circuit conditions.

Figure 12 shows the energy consumed in the reactor and the energy transfer efficiency from the primary capacitor to the reactor as a function of oxygen concentration in the reactor. The consumed energy i.e. the energy transfer efficiency is almost independent of the oxygen concentration in the circuit conditions 1-3. However, the consumed energy drastically decreases with increasing oxygen concentration at lower oxygen concentration than 40% under the circuit condition 4. When the oxygen concentration is larger than 40%, the consumed energy shows the constant value and is around 20 mJ. This value is almost the same to that of the circuit condition 3; *C*=0.48 nF and *L*=4.8 μH.

3.4 Influence of oxygen concentration on NO removal efficiency

The discharge properties in the corona reactor are changed by changing the value of the primary capacitor employed in the IES pulsed power supply and by the gas mixture such as oxygen concentration injected into the corona reactor as mentioned in Sec. 3.3. In this section, NO removal efficiency is described for various oxygen concentrations at two different primary capacitor capacitances to clarify the influence of the carrier gas mixture on NO removal efficiency.

Figure 13 shows the energy consumed in the reactor per one applied voltage pulse, the NO removal and its energy efficiency as a function of oxygen concentration in the NO contained

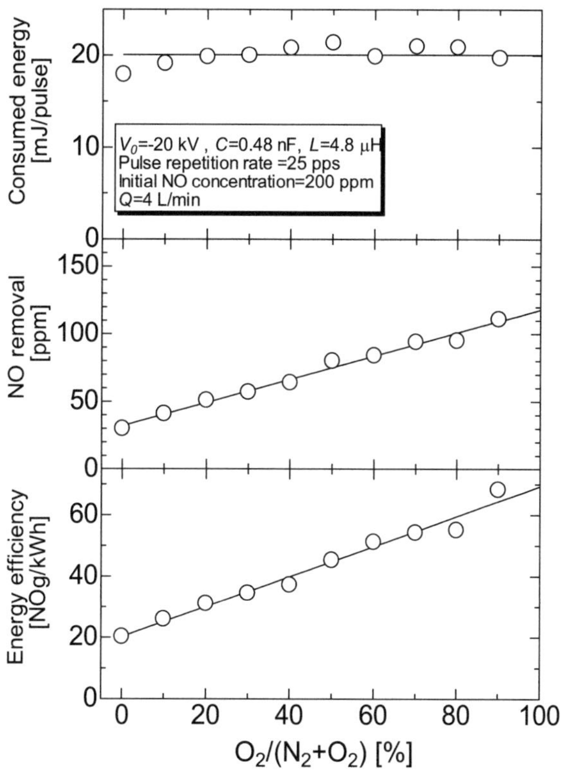

Fig. 13. Consumed energy, NO removal and its energy efficiency as a function of oxygen concentration in the NO contained gas mixture for the circuit condition 3; *C*=0.48 nF and *L*=4.8 μH at V_0= -20 kV, 25 pps in pulse repetition rate and 4 L/min in gas flow rate.

gas mixture under the circuit condition 3; C=0.48 nF and L=4.8 μH shown in Table 1. The charging voltage of the primary capacitor and the pulse repetition rate are -20 kV and 25 pps, respectively. The gas mixture flow into the reactor is 4 L/min. The consumed energy per pulse is almost constant and is 20 mJ. NO removal linearly increases from 30 to 110 ppm with increasing oxygen concentration from 0 to 90%. As a result, the energy efficiency for NO removal linearly increases from 20 to 65 g/kWh, which is calculated using equation (9), with increasing oxygen concentration from 0 to 90%. Generally, NO removal is caused via oxidization process as follows (Gentile and Kushner, 1995, Matzing, 1991):

$$NO + O + M \rightarrow NO_2 + M; \quad k_{11} = 9.0\times10^{-32}\, cm^6/s, \tag{11}$$

$$NO + O_3 \rightarrow NO_2 + O_2; \quad k_{12} = 1.7\times10^{-14}\, cm^6/s, \tag{12}$$

where M is the carrier gas concentration i.e. three-body reaction. The ozone and O radicals are produced via the following reactions:

$$e + O_2 \rightarrow O(^3P) + O(^3P) + e; \quad 6.1\ eV, \tag{13}$$

$$e + O_2 \rightarrow O(^3P) + O(^3D) + e; \quad 8.4\ eV, \tag{14}$$

$$O + O_2 + M \rightarrow O_3 + M; \quad k_{15} = 6.3\times10^{-34}\, cm^6/s. \tag{15}$$

The reaction rate of the reactions (13)-(15) increases with increasing oxygen concentration. As a result, the rate of the reactions (11) and (12) increases. This is one of the reasons for the increase of NO removal with oxygen concentration shown in Fig. 13.

Figure 14 shows the energy consumed in the reactor per one applied voltage pulse, NO removal and its energy efficiency as a function of oxygen concentration in the NO contained gas mixture under the circuit condition 4; C=4.2 nF and L=12.6 μH shown in Table 1. The charging voltage of the primary capacitor and the pulse repetition rate are -12 kV and 25 pps, respectively. The gas mixture is flow into the reactor with 4 L/min. The consumed energy per one pulse decreases from 200 to 36 mJ with increasing oxygen concentration from 0 to 90%. NO removal is increases from 35 to 150 ppm with increasing oxygen concentration from 0 to 90%. As a result, the energy efficiency for NO removal drastically increases from 2.1 to 51 g/kWh. In this circuit condition, the streamer-to-glow transition easily occurs in nitrogen rich conditions. However, the streamer-to-glow transition is hard to occur in oxygen rich conditions. As a result, NO can be removed with high energy efficiency in spite of the large capacitance of the primary capacitor.

In general, typical value of oxygen content in diesel engine exhaust gas is 10%, whereas the oxygen content changes in range from 16 to 6% by a load of the engine (Takaki et al., 2001). The energy efficiencies for NO removal under 10% oxygen content gas are 26 and 4.9 g/kWh for circuit conditions 3 and 4, respectively. Therefore, it is important to choose circuit parameter adequately in application for NO_x removal from diesel engine exhaust gas.

3.5 Prevention of glow generation

The experimental results show that the generation of the glow discharge lowers an energy yield for NO removal when the charge remains in the primary capacitor after opening the switch. The glow discharge is maintained with circuit current from the capacitor through the inductor as shown in Fig. 1. If the opening switch is connected with the discharge reactor in series, as shown in **Fig. 15**, the circuit current is interrupted and the discharge is quenched rapidly. Therefore, the NO removal experiment was carried out using a circuit with SOS diode in series connected to the reactor.

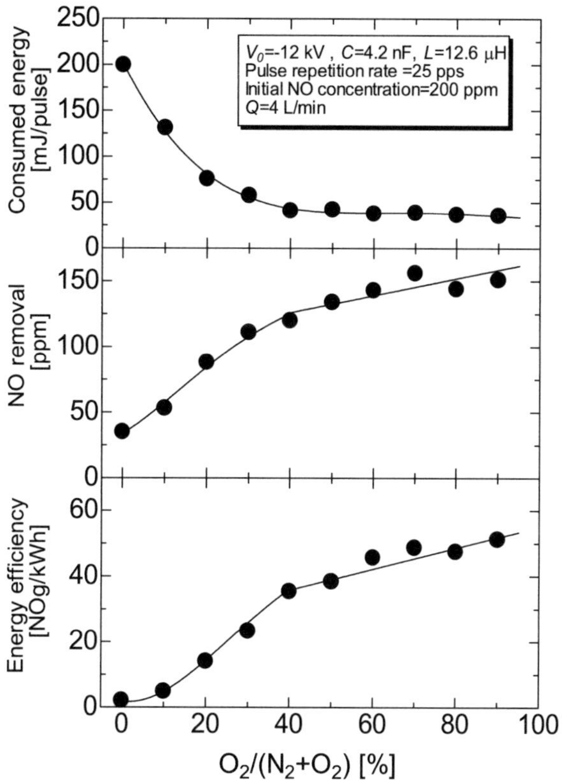

Fig. 14. Consumed energy, NO removal and its energy efficiency as a function of oxygen concentration in the NO contained gas mixture for the circuit condition; C=4.2 nF and L=12.6 μH at V_0= -12 kV, 25 pps in pulse repetition rate and 4 L/min in gas flow rate.

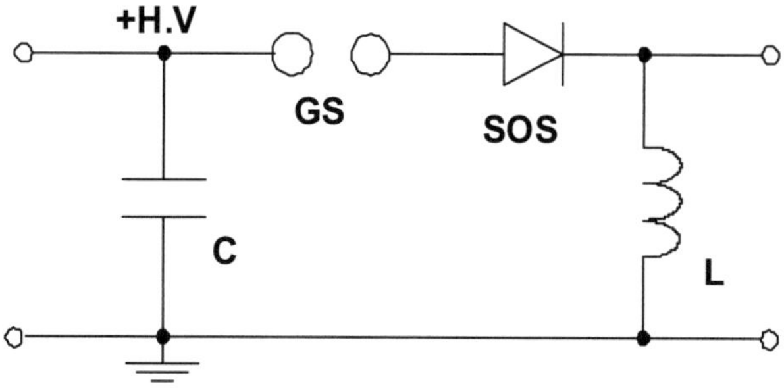

Fig. 15. Pulse power circuit for prevention of glow generation.

Figure 16 shows the waveforms of the discharge current, voltage between the electrodes, and the energy consumed for different SOS diode positions. The charging voltage of the capacitor is -12 kV. The symbol Cir1 represents the circuit connecting the SOS diode with the reactor in parallel, as shown in Fig. 1, and Cir2 represents the circuit connecting in series, as shown in Fig. 15. In Cir2, the voltage is applied to the reactor before opening the SOS because the inductor is connected to the reactor in parallel. However, the discharge current is almost zero before opening the SOS diode since the charging voltage of -12 kV is lower than the breakdown voltage of the discharge reactor. After opening the SOS diode, the pulse voltage over 20 kV is applied to the reactor; as a result, the discharge occurs, and the discharge current rapidly increases to 50 A. After that, the discharge current rapidly decreases within 50 ns without glow discharge generation. The energy consumed in the reactor of the Cir2 is about 15 mJ at 0.6 μs after opening the SOS diode, which is much less than the 120 mJ of Cir1.

Figure 17 shows NO removal efficiency and energy efficiency of NO removal as a function of input energy density in the reactor at various circuit conditions. The charging voltage in the primary capacitor is set to 12 kV. The LC condition I of 0.70 nF capacitor and 2.5 μH inductor ($\pi\sqrt{LC}/2 = 66$ ns) is employed to generate streamer discharge without a streamer-to-glow transition. The LC condition II is the same as condition 4 in which a streamer-to-glow transition occurs in Cir1. In the case of the LC condition I, NO removal efficiency for both circuits Cir1 and Cir2 is almost the same. However, in the case of the LC condition II,

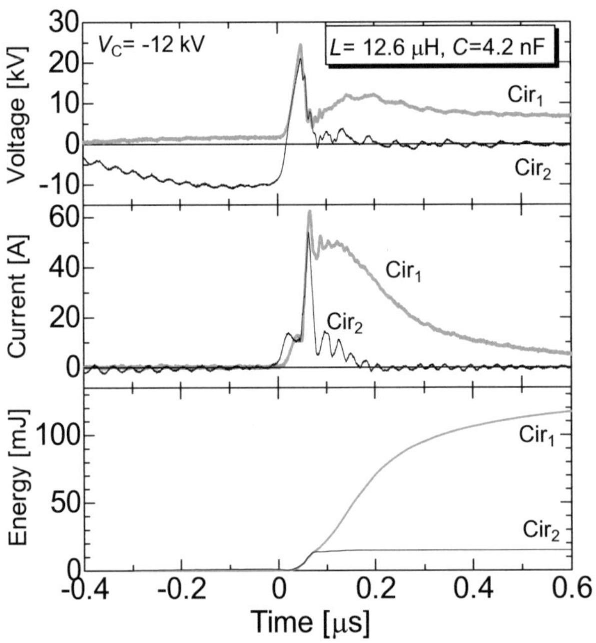

Fig. 16. Waveforms of applied voltage, discharge current, and energy consumed in the reactor for two different circuits.

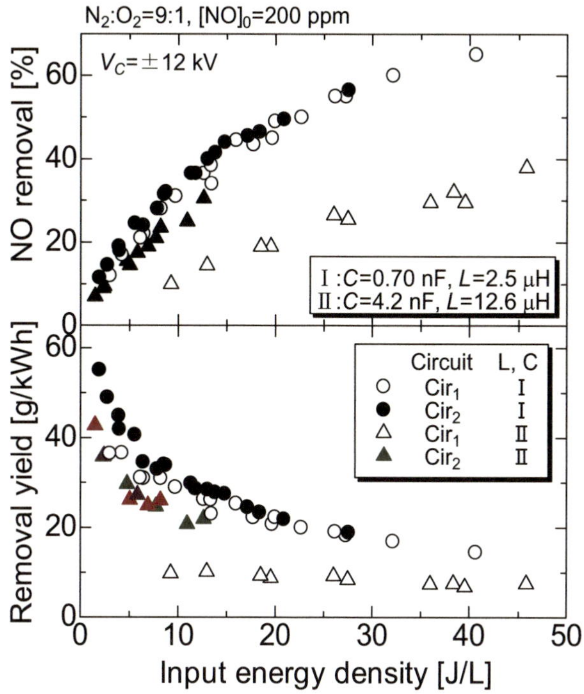

Fig. 17. Dependence of NO removal efficiency and energy efficiency on input energy density for two different circuits.

NO removal efficiency of Cir2 is much higher than that of Cir1. The arrangement of circuit 2, shown in Fig. 15, is effective for prevention of streamer-to-glow transition and to keep high energy efficiency for NO removal.

4. Conclusion

The experimental study on NO removal in a pulse corona discharge reactor was carried out to clarify the influence of the streamer-to-glow transition on NO removal. The inductive energy storage pulsed power generator was employed with a fast recovery diode as a semiconductor opening switch. The pulsed power generator supplied a 30 kV pulse with 300 pps repetition rate. The energy efficiency obtained for NO removal was 25 g/kWh at 30% removal. However, the energy efficiency decreased to 5 g/kWh with an increase in capacitance of the primary capacitor from several hundred pF to several nF. This decrease was caused by the streamer-to-glow transition. The efficiency was affected by oxygen concentration in the gas mixture. The efficiency was successfully improved from 5 to 20 g/kWh in the case of several nF-capacitors by changing the connection of the diode and the inductor in the pulsed power circuit.

5. References

Amirov, R.H., Asinovsky, E.I., Samoilov, I.S., and Shepelin, A.V. (1993) 'Oxidation characteristics of nitrogen monoxide by nanosecond pulse corona discharges in a methane combustion flue gas', *Plasma Sources Sci. Technol.*, Vol. 2, pp. 289–295.

Baksht, E. H., Panchenko, A. N., Tarasenko, V. F., Matsunaga, T. and Goto, T. (2002) 'Long-Pulse-Discharge XeF and KrF Lasers Pumped by a Generator with Inductive Energy Storage', *Jpn. J. Appl. Phys.*, Vol. 41, pp. 3701-3703.

Bushlyakov, A. I., Lyubutin, S. K., Ponomarev, A. V., Rukin, S. N., Slovikovsky, B. G., Timoshenkov, S. P. and Tsyranov, S. N. (2006) 'Solid-State SOS-Based Generator Providing a Peak Power of 4 GW', *IEEE Trans. Plasma Sci.*, Vol. 34, pp. 1873-1878.

Cathey, C.D., Tang, T., Shiraishi, T., Urushihara, T., Kuthi, A. and Gundersen, M. A. (2007) 'Nanosecond plasma ignition for improved performance of an internal combustion engine', *IEEE Trans. Plasma Sci.*, Vol. 35, pp. 1664-1668.

Chang, J. S., Lawless, P. A., and Yamamoto, T. (1991) 'Corona discharge processes', *IEEE Trans. Plasma Sci.*, Vol. 19, pp. 1152–1166.

Chang, J.S. (2001) 'Recent development of plasma pollution control technology: a critical review', *Sci. & Technol. Adv. Materials*, Vol. 2, pp. 571-576.

Clements, J.S., Mizuno, A., Finney, W.C. and Davis, R.H. (1989) 'Combined removal of SO_2, NOx, and fly ash from simulated flue gas using pulsed streamer corona', *IEEE Trans. Ind. Appl.*, Vol. 25, pp. 62-69.

Dinelli, G., Civitano, L. and Rea, M. (1990) 'Industrial experiments on pulse corona simultaneous removal of NO_x and SO_2 from flue gas', *IEEE Trans. Ind. Appl.*, Vol. 26, pp. 535–541.

Eliasson, B. and Kogelschatz, U. (1991) 'Nonequilibrium volume plasma chemical processing', *IEEE Trans. Plasma Sci.*, Vol. 19, pp. 1063-1077.

Gentile A. C. and Kushner, M. J. (1995) 'Reaction chemistry and optimization of plasma remediation of N_xO_y from gas streams', *J. Appl. Phys.*, Vol. 73, pp. 2074-2085.

Grekhov, I.V. and Mesyats, G.A. (2002) 'Physical basis for high-power semiconductor nanosecond opening switches', *IEEE Trans. Plasma Sci.*, Vol. 28, pp. 1540-1544.

Hackam, R. and Akiyama, H. (2000) 'Air pollution control by electrical discharges', *IEEE Trans. Diel. Elect. Insul.*, Vol. 7, pp. 654-683.

Hill, R. D., Rahmim, I. and Rinker, R. G. (1988) 'Experimental study of the production of NO, N_2O, and O_3 in a simulated atmospheric corona', *Ind. Eng. Chem. Res.*, Vol. 27, pp.1264-1269.

Jiang, W., Nakahiro, K., Yatsui, K., Kim, J. K. and Shimizu, N. (2007) 'Repetitive Pulsed High Voltage Generation Using Inductive Energy Storage with Static-induction Thyristor as Opening Switch', *IEEE Trans. Dielectr. Electr. Insul.*, Vol. 14, pp. 941-946.

Lowke, J. J. and Morrow, R. (1995) 'Theoretical analysis of removal of oxides of sulphur and nitrogen in pulsed operation of electrostatic precipitators', *IEEE Trans. Plasma Sci.*, Vol. 23, pp. 661-671.

Matzing, H. (1991) 'Chemical kinetics of flue gas cleaning by irradiation with electrons', *Adv. Chem. Phys.*, Vol. 80, pp. 315-402.

Nakamura Y. (1987) 'Drift velocity and longitudinal diffusion coefficient of electrons in nitrogen and carbon monoxide', *J. Phys. D: Appl. Phys.*, Vol. 20, pp. 933-938.

Namihira, T., Tsukamoto, S., Wang, D., Katsuki, S., Hackam, R., Akiyama, H., Uchida, Y. and Koike, M. (2000) 'Improvement of NO_x removal efficiency using short-width pulse power', *IEEE Trans. Plasma Sci.*, Vol. 28, pp. 434-442.

Penetrante, B.M., Hsiao, M.C., Merritt, B.T., Vogtlin, G.E. and Wallman, P.H. (1995) 'Comparison of electrical discharge techniques for nonthermal plasma processing of NO in N_2', *IEEE Trans. Plasma Sci.*, Vol. 23, pp. 679–687.

Robiscoe, R.T., Kadish, A. and Maier II, W.B. (1998) 'A lumped circuit model for transient arc discharge', *J. App. Phys.*, Vol. 64, pp. 4355-4363.

Rukin, S.N. (1999) 'High-power nanosecond pulse generators based on semiconductor opening switches (review) ', *Instruments and Experimental Techniques*, Vol. 42, pp. 439-467.

Takaki, K. and Akiyama, H. (1992) 'Development of compact pulsed power generator using inductive energy storage system', *Annual Report of Oita National College of Technology*, Vol. 28, pp. 19-24 (in Japanese).

Takaki, K., Kitamura, D. and Fujiwara, T. (2000) 'Characteristics of a high-current transient glow discharge in dry air', *J. Phys. D: Appl. Phys.*, Vol. 33, pp. 1369-1375.

Takaki, K. and Fujiwara, T. (2001) 'Multipoint Barrier Discharge Process for Removal of NO_x from Diesel Engine Exhaust', *IEEE Trans. Plasma Sci.*, Vol. 29, pp. 518-523.

Takaki, K., Shimizu, M., Mukaigawa, S. and Fujiwara, T. (2004) 'Effect of electrode shape in dielectric barrier discharge plasma reactor for NO_x removal', *IEEE Trans. Plasma Sci.*, Vol. 32, pp. 32-38.

Takaki, K., Hosokawa, M., Sasaki, T., Mukaigawa, S. and Fujiwara, T. (2005) 'Production of atmospheric-pressure glow discharge in nitrogen using needle-electrode', *Appl. Phys. Lett.*, Vol. 86, pp. 151501.1-3.

Takaki, K., Kanesawa, K., Mukaigawa, S., Fujiwara, T. and Go, T. (2007) 'Energy Efficiency of Corona Discharge Reactor Driven by Inductive Energy Storage System Pulsed Power Generator', *IEEE Trans. Dielectr. Electr. Insul.*, Vol. 14, pp. 834-845.

Vikharev, A.L., Gorbachev, A.M., Ivanov, O.A., Kolisko, A.L. and Litvak, A.G. (1993) 'Modeling of plasma chemical processes in the artificial ionized layer in the upper atmosphere by the nanosecond corona discharge', *Phys. Let. A*, Vol. 179, pp. 122-126.

Yalandin, M. I., Lyubutin, S. K., Oulmascoulov, M. R., Rukin, S. N., Shpak, V. G., Shunailov, S. A. and Slovikovsky, B. G. (2002) 'High peak power and high average power subnanosecond modulator opening at a repetition frequency of 3.5 kHz', *IEEE Trans. Plasma Sci.*, Vol. 30, pp. 1700-1704.

Yankelevich, Y. and Pokryvailo, A. (2002) 'High-power short-pulsed corona: investigation of electrical performance, SO_2 removal, and ozone generation', *IEEE Trans. Plasma Sci.*, Vol. 30, pp. 1975-1981.

Yankelevich, Y., Wolf, M., Baksht, R., Pokryvailo, A., Vinogradov, J., Rivin, B., and Sher, E., A. (2007) 'NOx diesel exhaust treatment using a pulsed corona discharge: the pulse repetition rate effect', *Plasma Sources Sci. Technol.*, Vol. 16, pp. 386-391.

7

Numerical Simulation on the Continuous Operation of Aquifer Thermal Energy Storage System

Kun Sang Lee
Kyonggi University
S. Korea

1. Introduction

As the demand for energy increases, any work to enhance energy conservation becomes crucial. Thermal energy storage (TES) system applications around the world have been known to provide economical and environmental solutions to the energy problems (Paksoy et al., 2004). TES systems contribute significantly to improving energy efficiency by helping match energy supply and demand. TES also makes it possible to more effectively utilize new renewable energy sources (solar, geothermal, ambient, etc.) and waste heat/cold recovery for space heating and cooling. With a storage medium of various types and sizes, TES systems therefore contribute to enhancing energy efficiency.

The storage medium can be located in containers of various types and sizes. Underground thermal energy storage (UTES) is mostly used for large quantities of seasonal heat/cold storage (Nielsen, 2003). There are several concepts as to how the underground can be used for underground thermal energy storage depending on geological, hydrogeological, and other site conditions. The two most promising options are storage in aquifers (aquifer thermal energy storage, ATES) and storage through borehole heat exchangers (borehole thermal energy storage, BTES) (Sanner[b], 2001; Andersson, 2007). In borehole thermal energy storage systems, also called "closed" systems, a fluid (water in most cases) is pumped through heat exchangers in the ground. In aquifer thermal energy storage or "open" systems, groundwater is pumped out of the ground and injected into the ground by using wells to carry the thermal energy into and out of an aquifer (Novo et al., 2010).

Aquifer thermal energy storage (ATES) system utilizes low-temperature geothermal resource in the aquifer (Sanner[a], 2001; Rafferty, 2003). Aquifer thermal energy storage, which is similar to the groundwater geothermal system under direct uses, involves storage and provides for both heating and cooling on a seasonal basis. An advantage of open systems is generally higher heat transfer capacity of a well compared to a borehole. This makes ATES usually the cheapest alternative if the subsurface is hydrogeologically and hydrochemically suited for the system. Such aquifers offer a potential and economical way of storing thermal energy for long periods of time. ATES systems have been used successfully around the world for the seasonal storage of heat and cold energy for the purpose of heating and cooling buildings (Probert et al., 1994; Paksoy et al., 2000; Allen et al., 2000; Schmidt, 2003; Paksoy et al., 2004).

Recently, the use of computer modeling has become standard practice in the prediction and evaluation of geothermal performance (Breger et al., 1996; O'Sullivan et al., 2001). In carrying out ATES development projects, a numerical modeling based on coupled mass and energy transport theory has to be conducted on the behavior of local subsurface geothermal system to evaluate and optimize a project design.
A number of researchers have highlighted the important role of numerical modeling in the analysis of ATES systems. Probert et al. (1994) presented the thermodynamic evaluations of ATES projects and listed key aquifer properties and design parameters. Based on an elementary ATES model, Rosen (1999) performed second-law analysis to thermal energy storage systems to assess overall system performance. Chevalier and Banton (1999) applied the random walk method of resolution to the study of energy transfer phenomena in ATES using a single injection well. Tenma et al. (2003) carried out a more realistic two-well model study to examine an underground design of TES system. To provide basic data for design, they evaluated the sensitivity of parameters affecting on the long-time performance of ATES. Their study, however, derived conclusions based on the simulation results for ATES systems under simple operation scenarios. From the preliminary simulations, the location of screen, which was extensively examined in the Tenma et al. (2003)'s model, is shown to have a relatively limited impact on the predicted temperature profiles of produced water. To overcome a limited applicability of previous researches, more comprehensive study should be established for the evaluation of the ATES systems. The present work extends previously reported researches to ATES systems under various operation parameters which are key factors influencing long-time performance of ATES systems. The main design considerations involve loading conditions including injection/production schedules and temperature, heat losses, and configuration of well-aquifer system simulating various design and operation scenarios.
Coupled hydrogeological-thermal simulations were undertaken in order to predict thermal behavior of aquifer and recovery temperatures from the aquifer. Analyses for a two-well system in a confined aquifer were performed in order to determine how operational parameters affect results of aquifer thermal energy storage simulations. The aim of the evaluation is to make reliable predictions about future recovery temperatures and temperature distributions in the aquifer given the planned injection/production temperatures and rates.

2. Basic concepts

Being similar to direct use of a groundwater-geothermal system, aquifer thermal energy storage involves drilling a few wells to an aquifer for circulation of water between the storage region and the energy system. Then it can store energy whilst providing heating and cooling on a seasonal basis. The wells are separated by a critical distance to ensure that the warm and cold storage remain separate and that thermal breakthrough does not occur within one season. This critical distance is primarily a function of operational and thermohydraulic parameters involving the well production rates, the aquifer thickness, and the hydraulic and thermal properties that control the storage volume. A plant can also be made with groups of wells instead of just two wells. Multiple-well configurations have been employed where large volumes of water are required and in systems where individual well yields are low. Single-well applications have also been employed using vertical separation of hot and cold groundwater where multiple aquifers exist.

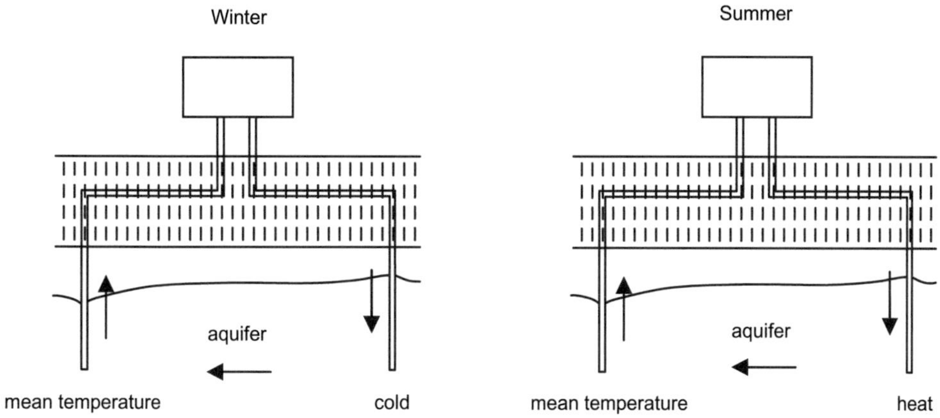

(a) continuous regime

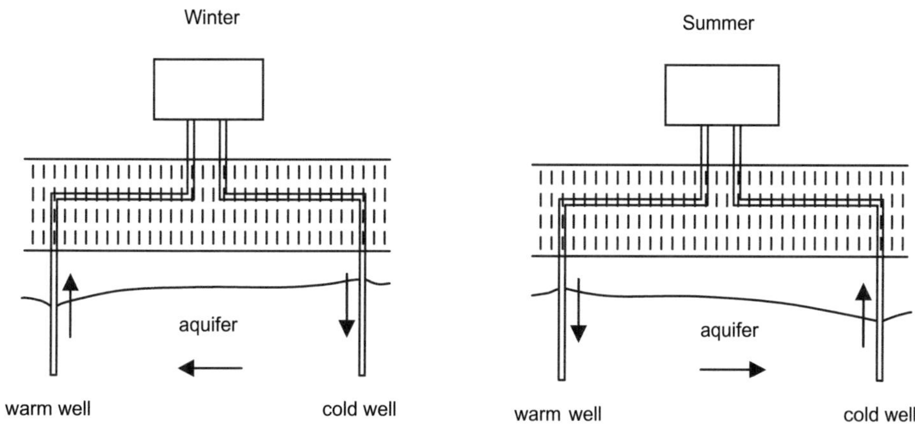

(b) cyclic regime

Fig. 1. Basic operational regimes for aquifer thermal energy storage (after Nielsen)

Usually, a pair of wells are pumped constantly in one direction or alternatively, from one well to the other, especially when both heating and cooling being provided. As presented in Figure 1, these two operation principles are called continuous regime and cyclic regime, respectively. The continuous regime only is feasible for plants where the load can be met with temperatures close to natural ground temperatures, and the storage part is more an enhanced recovery of natural ground temperatures. With a continuous flow, design and control of the system are much simpler and easier. Only one well or group of well needs to be equipped with pumps. Disadvantage is the limited temperature range. Cyclic flow will create a definite cold and heat reservoir around each well or group of wells. It is possible to maintain a ground volume above or below the natural ground temperature all the time. One disadvantage is a more complicated well design and control system with each well being able to both produce and inject groundwater.

3. Background of numerical simulation

3.1 Mathematical theory

The thermohydraulic analysis requires a calculation of groundwater flow and the temperature in the aquifer and the surrounding layers. In this section, theoretical principles of water flow and heat transfer phenomena for calculating temperatures of the aquifer at different locations were explained. The coupled groundwater and heat flow are governed by the partial differential equations describing mass and energy balance in the aquifer.
The conservation of mass for water in association with Darcy's law is expressed as continuity equation (Delshad et al., 1996; Clauser, 2003).

$$\frac{\partial}{\partial t}(n\rho_w)+\nabla\cdot(\rho_w \mathbf{u})=R_w \tag{1}$$

where
n: porosity
ρ_w: density of water
$\mathbf{u}$: Darcy flux
R_w: source/sink term
Specific discharge or Darcy velocity of water in the aquifer is defined by Darcy equation.

$$\mathbf{u}=-\frac{\mathbf{k}}{\mu_w}(\nabla p-\gamma_w \nabla z) \tag{2}$$

where $\mathbf{k}$ is the intrinsic permeability tensor; μ_w the viscosity; p the pressure; z the vertical depth, and γ_w the specific gravity (= $\rho_w g$).
The energy balance equation is derived by assuming that energy is a function of temperature only and energy flux in the aquifer occurs by convection and conduction only. The resulting general heat balance equation can be formulated as follows;

$$\frac{\partial T}{\partial t}[(1-n)\rho_s C_{vr}+n\rho_w C_{vw}]+\nabla\cdot(\rho_w C_{vw}\mathbf{u}T-\lambda_T\nabla T)=q_H-Q_L \tag{3}$$

where
T: aquifer temperature
C_{vr}, C_{vw}: rock and water heat capacity at constant volume
λ_T: thermal conductivity of aquifer
q_H: enthalpy source per unit bulk volume
Q_L: heat loss to overburden and underburden formations
The aquifer is assumed to be continuous and its thermal conductivity and volumetric heat capacity are considered to be a function of porosity and the thermal characteristics of water and soil matrix (Nassar et al., 2006). Thermal dependence of density, viscosity, thermal conductivity, and heat capacity is not taken into consideration because these parameters vary little in the considered temperature range.
The heat transfer with over and underlying low-permeability layers, is assumed to be due solely to thermal diffusion. The heat loss is computed by using the Vinsome & Westerveld (1980)'s method.

$$Q_L=\nabla\cdot(\lambda_{Te}\nabla T) \tag{4}$$

where λ_{Te} is the thermal conductivity of overburden or underburden rock. The heat transfer equation (3), which results from the principle of energy conservation, is coupled with the flow equation from Darcy's law (2) and the continuity equation (1).

3.2 Simulation model

Groundwater flow and thermal energy transport in the porous media have been studied in detail in the discipline of hydrogeology. Numerical research into groundwater and heat transport has been continuing for more than a decade in North American and Europe. Numerous commercially available and public domain numerical software codes exist. Of these, focus is given to the simulation modeling both mass and heat transport in groundwater.

Many simulation codes available to simulate ATES systems have their own merit. Deng (2004) and Schmidt & Hellström (2005) give a summary of available numerical models for groundwater flow and energy or solute transport in groundwater. These models can all be used to simulate an ATES system.

Amongst the more sophisticated simulators, a general simulator named UTCHEM has proved to be particularly useful for modeling multiphase transport processes under non-isothermal condition (Center for Petroleum and Geosystems Engineering, 2000). The simulator was originally developed to simulate the enhanced recovery of oil using surfactant and polymer processes. UTCHEM has been verified by comparing its ability to predict the flow of fluids through the aquifer to analytical solutions and experimental measurements.

4. Numerical modeling

An understanding of the thermohydraulic processes in the aquifer is necessary for a proper design of an ATES system under given thermohydraulic conditions. The main design considerations concern the loading conditions (injection temperature and pumping rates), heat losses, thermal breakthrough, etc (Claesson et al., 1994). The model gives the temperature of the produced water and groundwater in the aquifer.

A multidimensional, finite-difference model for ground-water flow and heat transport is used to solve the complex thermohydraulic problems numerically. The numerical model includes the effects of hydraulic anisotropy, thermal convection and conduction, and heat loss to the adjacent confining strata. .In order to decide the sustainability of the aquifer for energy storage application, temperatures of producing water is estimated over a ten-year period.

In order to estimate the parameters of an underground system, a general ATES using the open system of 5-spot patterns is considered. As shown in Fig 2(a), four wells are situated at the corners of a square field. The fifth well is located at the center of square. Water is pumped into the injection well at a constant flow rate Q at a temperature Tinj, and the same flow rate of water is recovered from the neighboring four production wells or vice versa. In a large reservoir with repeated patterns, the flow is symmetric around each injection well with $0.25Q$ from each well confined to the pattern.

The 5-spot pattern can be further simplified to a two-well system because each quadrant is symmetric. A model of two-well on either corner with hydraulic coupling as shown in Fig. 2(b), has been developed to estimate the performance of symmetric array of wells in 5-spot pattern.

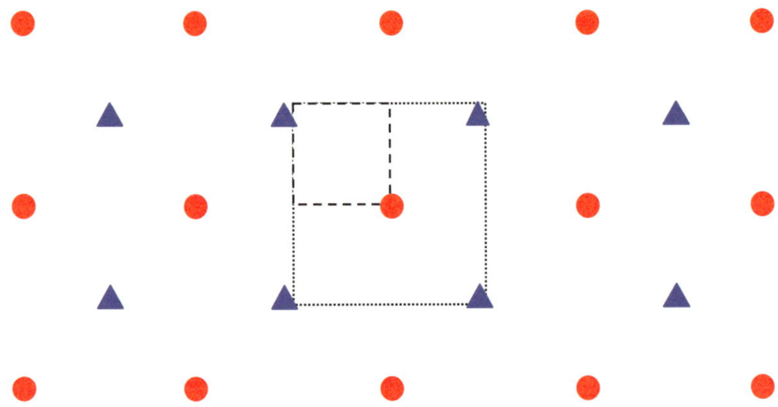

(a) 5-spot pattern

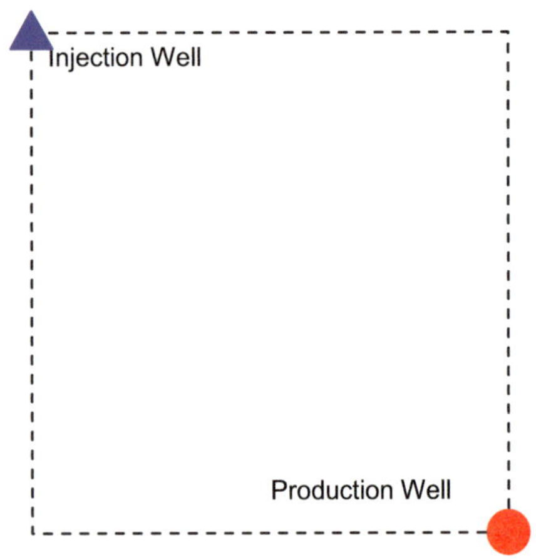

(b) two-well model for quadrant of a 5-spot pattern

Fig. 2. Array of wells in a 5-spot pattern

The outer boundary is represented as a noflow and adiabatic boundary to simulate symmetry in an array.

$$(\nabla T) \cdot \mathbf{n} = 0 \tag{5}$$

The model is similar to the model suggested by Tenma et al. (2003). This model is a unit of the system and composed of two wells partially-penetrating a confined aquifer. As the grid is divided into 13 grid blocks in horizontal directions, and 11 grid blocks in a vertical direction, there are total of 3,718 grid blocks. Well depth was set at the relatively shallow value of for the two-well system. Water level in the model was set at 6 m.

Determination of the potential of a specific confined aquifer as an effective thermal energy storage medium requires thorough knowledge of thermodynamic and hydraulic properties of the aquifer and its confining layers and fluid properties. Parameters include porosity and permeability of the storage aquifer and thermal conductivities and heat capacities of the aquifer matrix, native ground water, and confining layers. As presented in Table 1, the aquifer and water have constant thermal properties and assumed to be slightly compressible. The volumetric heat capacity and thermal conductivity were identical for the aquifer and the confining layers.

aquifer	porosity (n)	0.40
	permeability (k)	1,013 md
	compressibility of formation (β_r)	2.96×10^{-6} kPa^{-1}
	density of rock (ρ_r)	2.65 g/cm^3
	thermal conductivity of rock (λ_T)	249.2 kJ/day·m·°K
	thermal conductivity of overburden/underburden rock (λ_{Te})	249.2 kJ/day·m·°K
	heat capacity of rock (C_{vr})	0.8864 kJ/kg·°K
water	viscosity (μ_w)	1.1404 cp
	compressibility (β_w)	4.4×10^{-7} kPa^{-1}
	density (ρ_w)	1 g/cm^3
	heat capacity (C_{vw})	4.184 kJ/kg·°K

Table 1. Hydrogeological and thermal properties of aquifer and water

5. Results and discussion

To estimate the characteristics of the system, this two-well model was run for continuous flow regime. Heat transfer within the aquifer is simulated by specifying constant temperature T_{inj} at the injection well and with the aquifer temperature initialized at T_i. The initial temperature of the aquifer is assumed to be constant 17.5°C over the entire aquifer and confining layers. The model is run for 10 years to provide an adequate long-term assessment of thermal storage.

In the continuous flow regime, water is pumped from one well equipped with a pump and injected through a second well. A complete energy storage cycle is composed of four periods per year to simulate the seasonal conditions. Each cycle is symmetrical for identical injection ($Q > 0$) and production ($Q < 0$) rates and duration. The thermal field and the temperature of produced water, T_{prod}, are calculated from the numerical solutions of Equation (3) at constant time interval.

In order to evaluate and compare results obtained from ATES system simulations, a measure of performance is required. There is no generally valid basis for comparing the achieved performance of TES system under different conditions. In this study, the ratio of differential energy returned from aquifer to the energy originally in the aquifer, which is similar to dimensionless parameter adopted by Chavalier & Banton (1999) or Tenma et al. (2003), is used to measure TES performance.

$$\text{normalized thermal storage} = \frac{T_{prod} - T_i}{T_i} \tag{6}$$

A number of factors may influence the performance of the ATES. In this study, the following parameters were varied.

1. operation schedule: continuous and variations of cyclic regimes
2. heat loss to adjacent layers

3. temperature of injecting water
4. flow rate of water
5. distance between injection and production wells
6. aquifer thickness
7. screen length
8. anisotropy in vertical permeability

5.1 Operation schedule

The influence of a number of operation schedules on the performance of ATES was examined. The seasonal variations in the surface due to natural seasonal variations are involved in the operation schedules. Four cases of scenarios were considered in the present simulations, as list in Table 2.

	Injecting or Producing Flow Rate (m^3/day)				Temperature of Injecting Water(°C)			
Month / Case	1-3	4-6	7-9	10-12	1-3	4-6	7-9	10-12
Case 1	50	50	50	50	5	15	25	15
Case 2	50	50	50	50	5	25	5	25
Case 3	50	50	50	50	5	5	25	25
Case 4	50	0	50	0	5	-	25	-

Table 2. Description of continuous operation scenarios for a year

Accounting for the seasonal changes in surface temperature, the injected water temperatures in Case 1 were taken as 5°C, 15°C, 25°C, and 15°C over a three-month period, respectively. Case 1 will be used as a base case for other simulations. Operation schedules of Cases 2 and 3 were scenarios presented by Tenma et al. (2003). In these cases, the inlet temperature was held constant at 5°C for a half year, then changed to a temperature of 25°C for the rest of the year. Case 4 is similar to Case 1, except rest periods of neither injection nor production between 5°C and 25°C water injection periods. During the rest periods in Case 4, temperature of a cell containing production well is considered as T_{prod}. The size of this two-well model is 39 m × 39 m × 22 m. The rates of injection or production of 50 m^3/day correspond to 0.45094 pore volume of the aquifer for three months.

The evaluation of a TES system was performed by the temperature of produced water shown in Fig. 3. Reflecting the changes in temperature of the injecting water, the recovery temperatures fluctuate with a quarterly year period. Balances between injecting and producing thermal energy and small variation in temperatures are the most desirable case, representing sustainable use of ground as an ATES system without a negative effect on the environment.

As the temperature of the circulated fluid changes by 25-5°C, the temperature eventually becomes 15°C, the average temperature of injected water. Due to energy imbalance between the initial temperature of 17.5°C and average injecting temperature of 15°C, the aquifer undergoes a gradual cooling process and negative value of balance of thermal energy. The temperatures at the producing well were constant within 2.6°C for Cases 2 and 4 and 7.8°C for Case 3 through the ten-year period.

The results of normalized thermal balance are shown in Fig. 4. A positive value means the energy produced is larger than the initial energy; a negative value means the initial energy

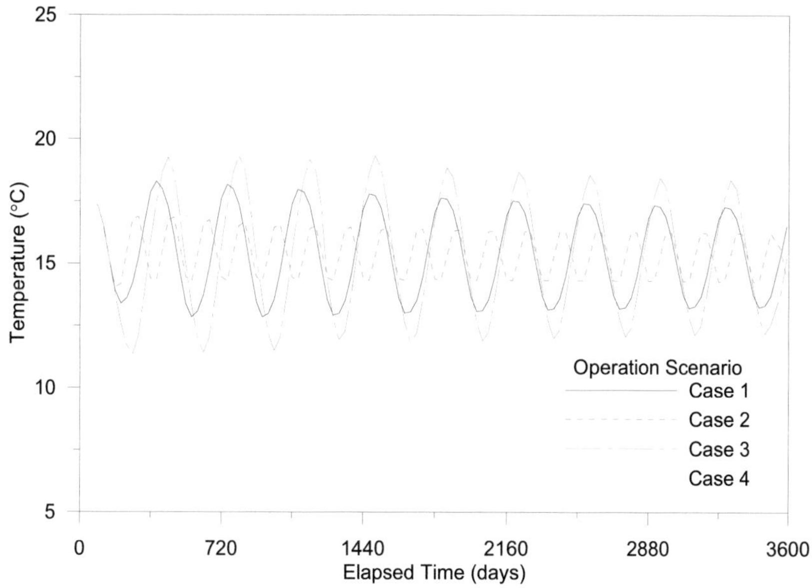

Fig. 3. History of recovery temperature obtained from simulations with different operation scenarios

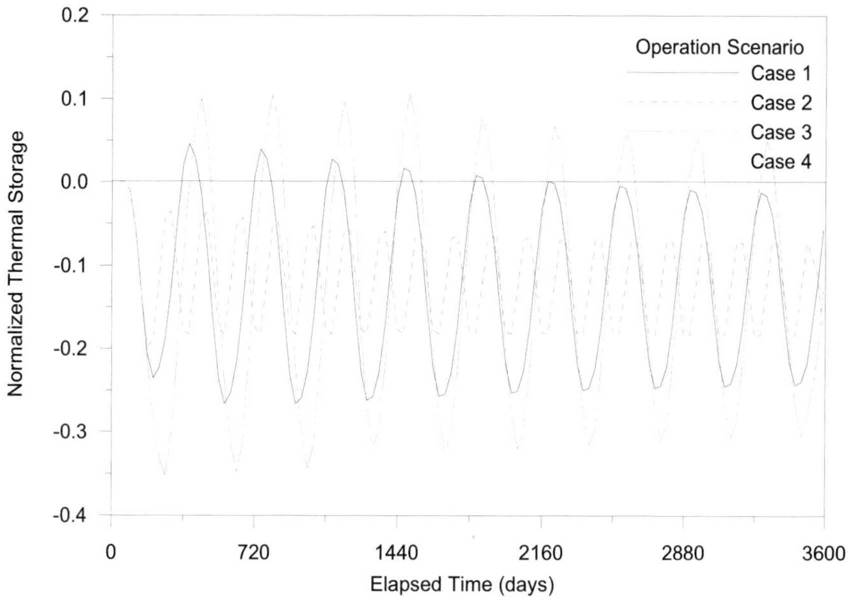

Fig. 4. History of balance of thermal energy obtained from simulations with different operation scenarios

is larger than the energy produced. Zero mean balance of thermal energy and small variation are the most desirable case, representing sustainable use of aquifer as a TES system. As shown in the Figure, the balance of thermal energy gradually decreases in all the cases. Reflecting the changes in temperature of the injecting water, the balance of thermal energy also fluctuates with a quarterly year period. The sum of normalized thermal storage values is smallest for Case 4 and largest for Case 3. Differences among cases are relatively small, less than 20%. However, the range of variation of Case 3 is the highest and 2.7 times higher than that of Case 4 which is the smallest. As the net change of thermal energy is small, the flow conditions of Cases 2 and 4 are promising.

One of objectives of numerical simulation is to visualize the groundwater flow and heat transfer in the aquifer and the movement of thermal front. With results from numerical simulations, one can provide graphical presentation of the groundwater flow and the motion of thermal fronts for given set of wells-aquifer configuration. The groundwater flow around the wells takes place mainly in the radial direction. The interface or thermal front is between the injected water and the water in the aquifer. The location of thermal front is determined by inspecting temperature distribution.

For the graphical representation of results, an illustrative example is taken from Case 1 for heat and cold storage. In Figs. 5(a) to (d), the simulated temperature distributions are presented with different colors for different temperatures. The temperatures are for the middle layer obtained from numerical calculations after 90, 180, 270, and 360 days of operation. These give direct pictures on the characteristics of energy flow fields and the location of thermal front. In the figures, there are clear evidences for energy storage. The area of relatively uniform temperature near the injection well represents the region influenced by water injected during each flow period. Clearly shown in the aquifer during winter and summer periods, the thermal fronts are from cold and warm water injected. They are located at a considerable distance from the production well. The pair of wells interacts slightly and the production well seems undisturbed by the injection well. Whilst the temperature of the injecting water changes by 25-5°C, the temperature of the produced water lies within much smaller range.

5.2 Heat loss

Heat transfer from/to overburden and underburden formations is considered as an important factor that may control the temperature of produced water. In the present work, results from the base case were compared with those from a case in which heat loss is not included. These formations are assumed to have the same thermal properties with aquifer, as stated earlier.

Figure 6 shows these results at different times. With heat loss, the average and range of thermal balance decreases gradually over time. Contrarily, the temperature at the production well fluctuates in the fixed range between 11.3°C and 19.6°C without heat loss. Therefore, the difference between results from two cases is gradually increasing. The range of thermal balance also remains constant at 0.53 which is higher than that obtained from the base case. The results indicate that conductive heat exchange with the surrounding rock is an important process causing different cycle of temperature variations in the production well.

5.3 Injection temperature

To demonstrate the effect of temperature difference on long-time thermal storage performance of the aquifer, additional simulations were performed using different combinations of water temperature at the injection side. Having average value of 15°C, the temperature of injecting water was constant during each three-month injection period and had values of 1°C/15°C/29°C/15°C and 9°C/15°C/21°C/15°C, respectively. Other conditions and parameters were not changed from the operation scenario of Case 1 with 5°C/15°C/25°C/15°C.

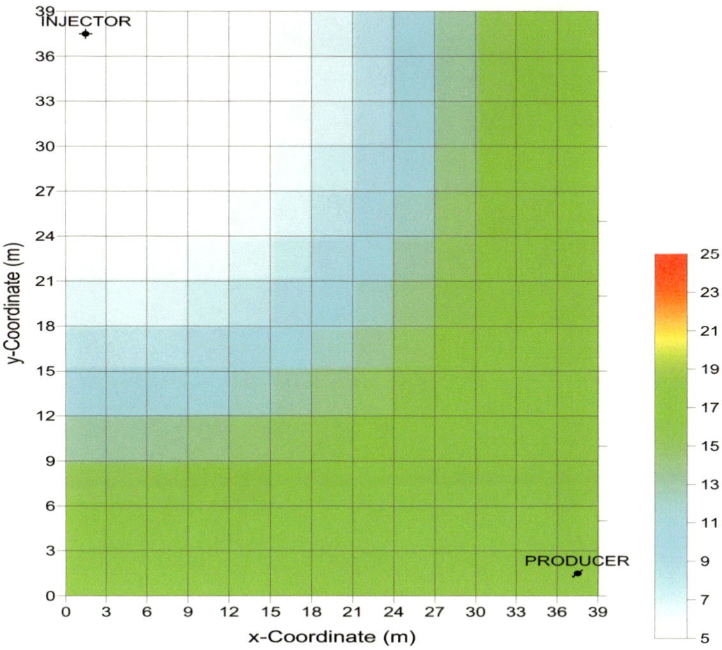

(a) 90 days

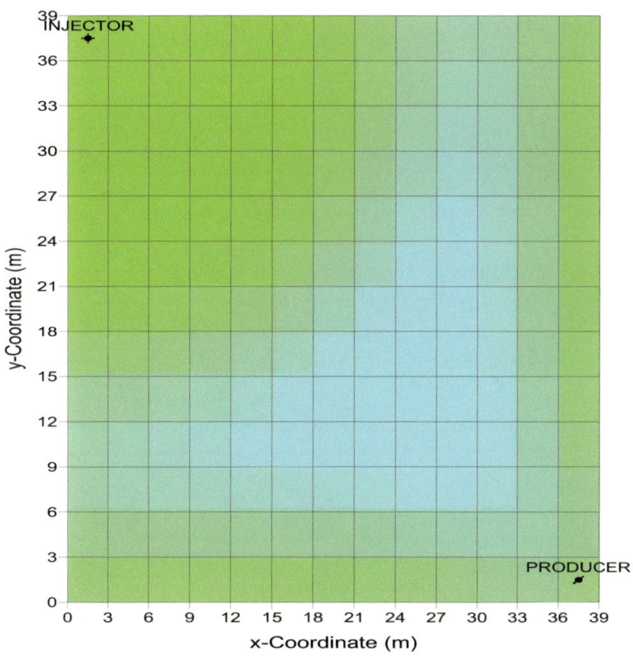

(b) 180 days

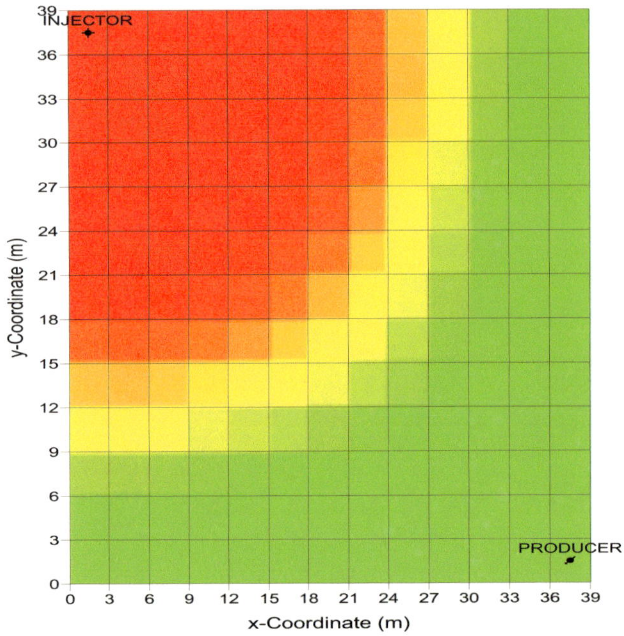

(c) 270 days

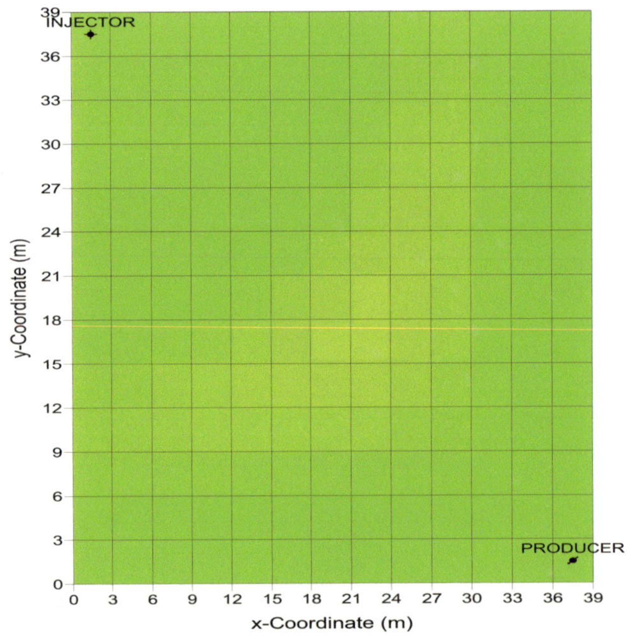

(d) 360 days

Fig. 5. Temperature distribution [°C] of the middle layer after cold/warm water injection during the first year

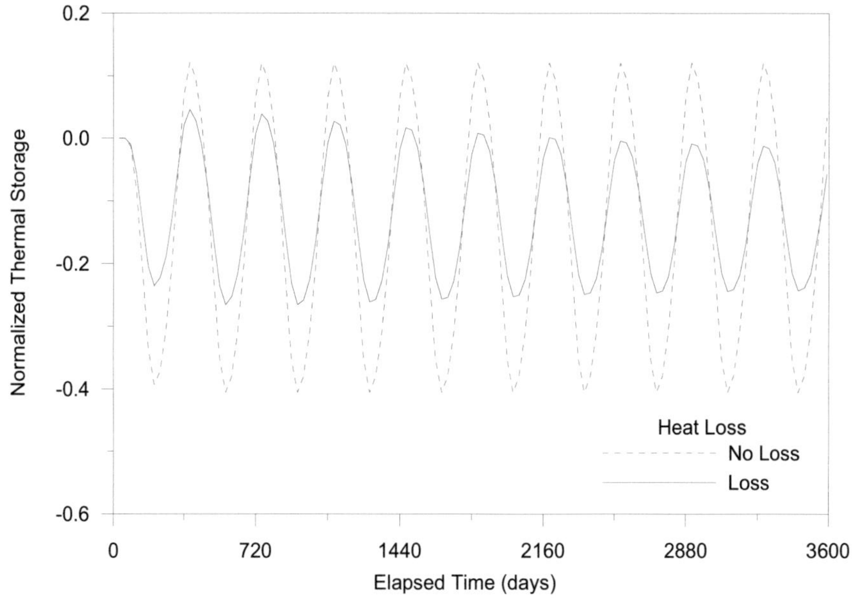

Fig. 6. History of balance of thermal energy obtained from simulations with different heat loss conditions

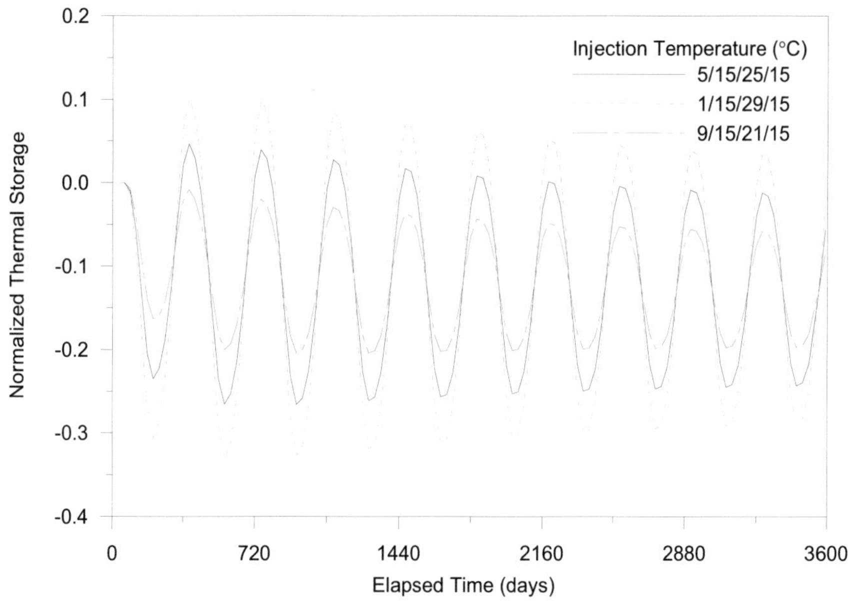

Fig. 7. History of balance of thermal energy obtained from simulations with different injection temperatures

The profiles in Fig. 7 shows increasing range in thermal balance with increasing differences in injection temperature, in accordance with the gradual drop over time. The range of variation of Case 1°C/15°C/29°C/15°C is the highest and 2.1 times higher than that of 9°C/15°C/21°C/15°C which is the smallest. By increasing the temperature difference around the year at the injection well from 12°C to 28°C, the normalized thermal storage is proportionally decreased. The sum of values is smallest for 9°C/15°C/21°C/15°C and largest for 1°C/15°C/29°C/15°C. Differences among summed values are relatively small, less than 20%.

5.4 Flow rate

The aim of this simulation is to evaluate the recovery of thermal energy from the aquifer given injection or production flow rates. The calculations were performed for well and aquifer configuration which is the same as the base case. However, in this simulation, the flow rates of injection and production are changed to 25, 50, and 75 m^3/day. These rates are equivalent to injecting 0.22547, 0.45094, and 0.67641 pore volumes for a three-month period, respectively.

Figure 8 presents a considerable drop in thermal storage and a significant increase in variation of the values with increasing flow rate. Tripling the flow rate results in increases in the sum of thermal storage values by 36% and ranges by 2.5 times. The result suggests that, everything else being the same, the use of less flow rate is a promising flow condition due to small net change in thermal energy. However, decreasing the flow rate considerably less than an appropriate value would not be effective because the resulting decrease in the amount of available thermal energy would be a problem.

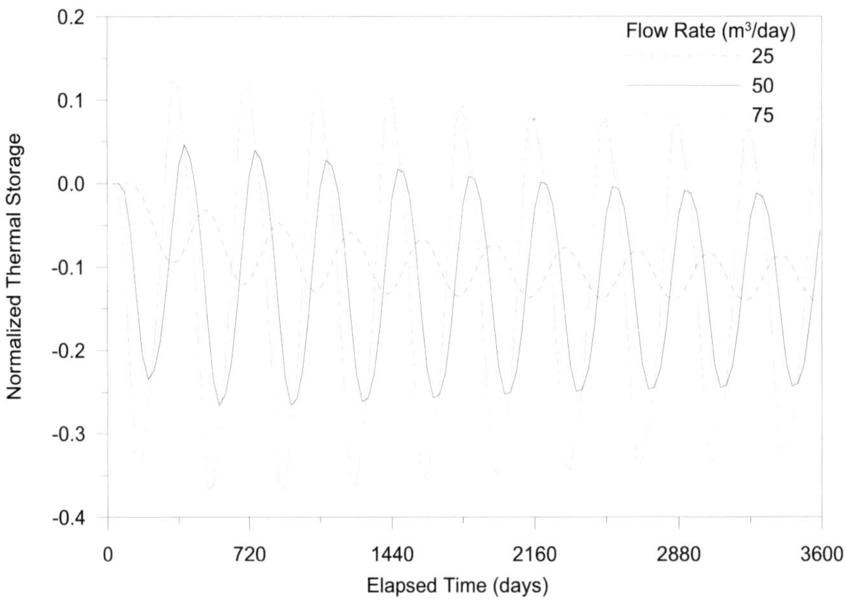

Fig. 8. History of balance of thermal energy obtained from simulations with different injection/production flow rates

5.5 Well distance

The distance between the injection and production wells influence the proportion of the aquifer that is effective in the heat transfer and thermal storage process. One objective with the numerical simulations is to find the well configuration to make the energy storage as dense as possible. In attempts to investigate effects of interwell distance, pumping and injection of groundwater were simulated for three cases where the sizes of computation domains are 32.5×32.5 m^2, 39.0×39.0 m^2, and 45.5×45.5 m^2. The injection and production wells are located 42.2 m, 50.9 m, and 59.4 m apart, respectively. These distances correspond to 0.64935, 0.45094, and 0.33130 pore volume of the aquifer for three months of injection or production at 50 m^3/day.

Figure 9 illustrates the energy balance for different well distances. A longer well distance reduced the variations in thermal storage substantially. Although almost doubling the pore volume of the aquifer by increasing the well-to-well distance does not double the performance of ATES systems, it does improve the performance of ATES systems by decreasing the sum of thermal storage values by 14% and ranges by 55%. Larger variations for shorter well-to-well distance result from the fact that the thermal front of injected water approaches the production well within each operation period. Therefore, the region near a producing well is considerably affected by injected water. This observation emphasizes the importance of ensuring that adequate distance between wells is used, taking into account the thermal and hydraulic transport of injected water.

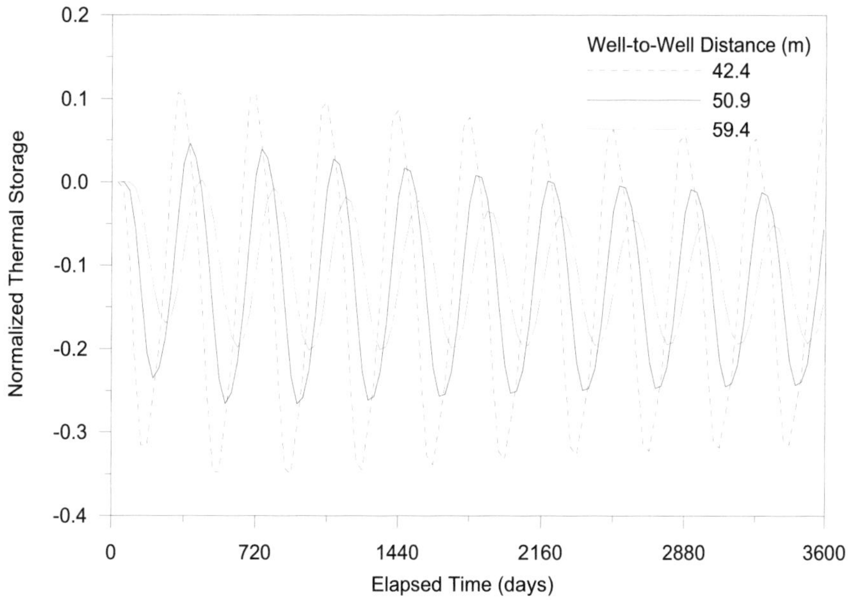

Fig. 9. History of balance of thermal energy obtained from simulations with different interwell distances

5.6 Aquifer thickness

To investigate the influence of the aquifer thickness in the performance of ATES system, two-well models with 22 m, 30 m, and 38 m thick aquifers were analyzed. The pore volumes of aquifers correspond to 0.45094, 0.33069, and 0.26107 for three months of injection or production at 50 m^3/day. Screen is 6 m long and installed at the center of the aquifer for all cases.

Figure 10 shows that increasing the thickness of the aquifer from 22 m to 38 m decreases the range of thermal storage by 32%, but the sum of thermal storage only by 3%. Although the results demonstrate the importance of aquifer thickness, they also show that thick aquifers would not be so effective in thermal energy storage. Compared with results from the simulation with different interwell distance and similar pore volume, the resulting decrease of the net changes in thermal energy is limited.

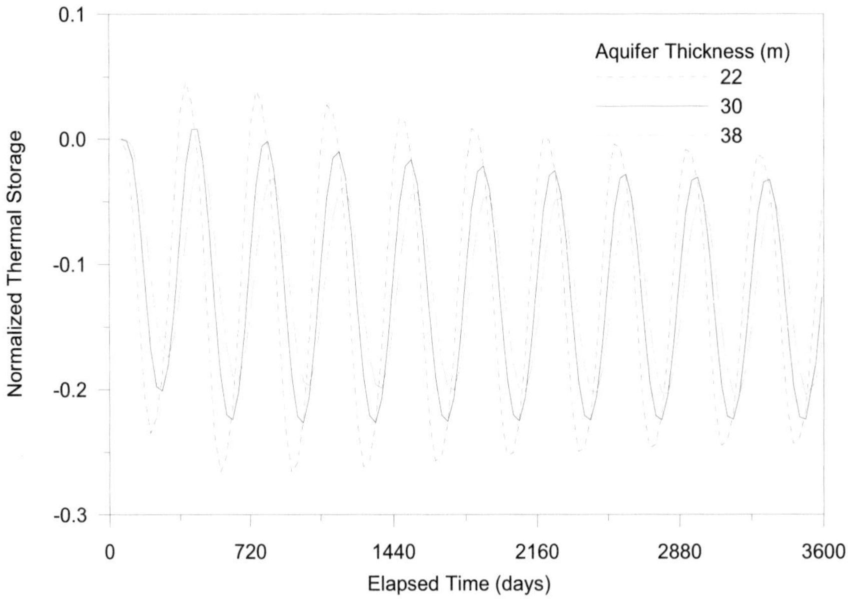

Fig. 10. History of balance of thermal energy obtained from simulations with different aquifer thicknesses

5.7 Screen length

To investigate how the screen length affects the performance of ATES system, two-well models is applied for aquifer of 39×39×22 m^3 with injection or production at 50 m^3/day. Screens for injection wells are 6 m long and installed at the center of the aquifer for all cases. For production wells, screens are 2m, 6 m, and 10 m long.

Figure 11 shows that increasing the length of the screen from 2 m to 10 m makes no differences in the range of thermal storage. The results show that long screen would not be so effective in thermal energy storage as long as vertical permeability is as high as horizontal permeability.

5.8 Vertical permeability

In this simulation, the effect of directional permeability on the performance of an ATES system was examined. All other conditions are similar to the base case except for considering anisotropy in permeability by replacing vertical permeability with different values. Permeabilities in x and y directions are kept constant. The ratios of vertical to horizontal permeability considered in this study are 1.0, 0.5, and 0.1.

Figure 12 presents the influence of vertical permeability on the performance of ATES systems in terms of values and variations of thermal storage. A weak dependence on the permeability anisotropy is obtained. The thermal storage values for k_z/k_x = 0.5 and

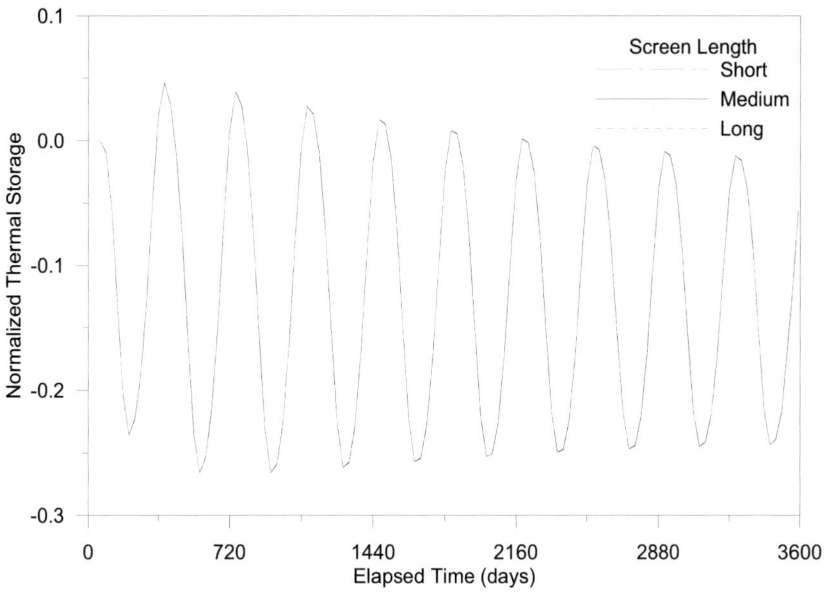

Fig. 11. History of balance of thermal energy obtained from simulations with different screen length

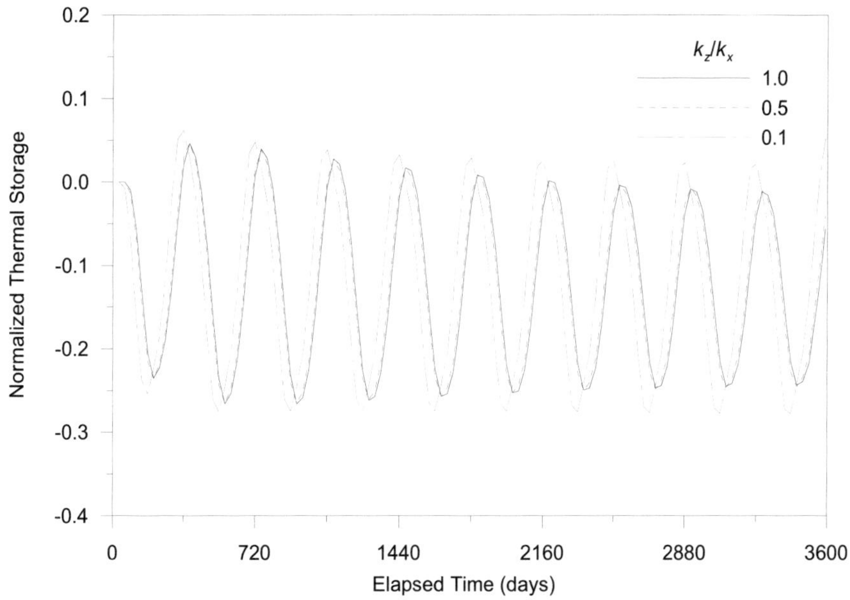

Fig. 12. History of balance of thermal energy obtained from simulations with different permeability anisotropy

k_z/k_x = 1.0 are almost identical, while those of k_z/k_x = 0.1 show a larger variations and less drop over time. However, the increase in the range of thermal storage is less than 9% and the sum of thermal storage only by 2%. The results suggest that, everything else being the same, the effects of permeability anisotropy would be only marginal.

6. Conclusion

A mathematical model describing the water flow and convective/conductive thermal energy transport in an ATES system is presented. The three-dimensional thermal process with combined groundwater and heat flow in the aquifer and heat conduction in surrounding layers is solved numerically. This paper presents the results of long-time thermal behavior of ATES system with two wells under continuous operation methods. The effects of various injection-withdrawal rates and durations on computed values of aquifer thermal behavior and final producing temperature were studied for a 10-year continuous injection and withdrawal. The hypothetical simulations indicate that the model of the two-well system will be a valuable tool in determining the most efficient system operation.
The thermal behavior of the storage system is shown to depend on the aquifer's volume relative to energy input and flow pattern of the water. Various operational and geometrical parameters including operation schedules, injection temperature, injection/production rates, and geometrical configuration of well and aquifer impact the predicted recovery water temperature. Small variations in injection temperatures, low flow rate, and large surface to volume ratio are recommended as an effective ATES because of small loss and little fluctuation in extracted thermal energy. However, aquifer thickness and hydraulic anisotropy have a minimal effect on the performance of ATES systems.

7. References

Allen D.M.; Ghomshei, M.M.; Sadler-Brown, T.L.; Dakin, A. & Holtz, D. (2000). The current status of geothermal exploration and development in Canada, *Proceedings of World Geothermal Congress 2000*, pp. 55-58, Kyushu-Tohoku, Japan, May 28-June 10 2000

Andersson, O. (2007). Aquifer thermal energy storage, In *Thermal Energy Storage for Sustainable Energy Consumption*, Paksoy, H.O. (Ed.), 155-176, Springer: Dordrecht, The Netherlands

Breger, D.B.; Hubbell, J.E.; Hasnaoui, H.E. & Sunderland, J.E. (1996). Thermal energy storage in the ground: comparative analysis of heat transfer modeling using U-tubes and boreholes, *Solar Energy*, Vol. 56, No. 6 (June 1996) 493-503

Center for Petroleum and Geosystems Engineering, (2000). *UTCHEM-9.0 A Three-Dimensional Chemical Flood Simulator*, University of Texas at Austin: Austin, TX, U.S.A.

Chavalier, S. & Banton, B. (1999). Modelling of heat transfer with the random walk method. Part 1. Application to thermal storage in porous aquifers, *Journal of Hydrology*, Vol. 222, No. 1-4 (September 1999) 129-139

Claesson, J.; Hellström, G. & Probert, T. (1994). Simulation models for ATES, *Proceedings of International Symposium on Aquifer Thermal Energy Storage,* pp. 131-137, Tuscaloosa, Alabama, U.S.A., November 14-15 1994

Clauser, C. (2003). *Numerical Simulation of Reactive Flow in Hot Aquifers*, Springer-Verlag: Berlin, Germany

Delshad, M.; Pope, G.A. & Sepehrnoori, K. (1996). A compositional simulator for modeling surfactant enhanced aquifer remediation, 1 fomulation, *Journal of Contaminant Hydrology*, Vol. 23, No. 4, (August 1996) 303-327

Deng, Z. (2004). *Modeling of Standing Column Wells in Ground Source Heat Pump Systems,* Ph.D. dissertation, Oklahoma State University: Stillwater, OK, U.S.A.

Nassar, Y.; ElNoaman, A.; Abutaima, A.; Yousif, S. & Salem, A. (2006). Evaluation of the underground soil thermal storage properties in Libya, *Renewable Energy*, Vol. 31, No. 5, (April 2006) 593-598

Nielsen, K. (2003). *Thermal Energy Storage, A State-of-the-Art*, Norwegian University of Science and Technology (NTNU): Trondheim, Norway

Novo, A.V.; Bayon, J.R.; Castro-Fresno, D. & Rodriguez-Hernandez, J. (2010) Review of seasonal heat storage in large basins: Water tanks and gravel-water pits, *Applied Energy*, Vol. 87, No. 2, (February 2010) 390-397

O'Sullivan, M.J.; K. Pruess & Lippmann, M.J. (2001). State of the art of geothermal reservoir simulation, *Geothermics,* Vol. 30, No. 4, (August 2001) 395-429

Paksoy, H.O.; Andersson, O.; Abaci, H.; Evliya, H. & Turgut, B. (2000). Heating and cooling of a hospital using solar energy coupled with seasonal thermal energy storage in aquifer, *Renewable Energy*, Vol. 19, No. 1-2, (January 2000) 177-122

Paksoy, H.O.; Gurbuz, Z.; Turgut, B.; Dikici, D. & Evliya, H. (2004). Aquifer thermal storage (ATES) for air-conditioning of supermarket in Turkey, *Renewable Energy*, Vol. 29, No. 12, (October 2004) 1991-1996

Probert T.; Hellsröm, G. & Glaesson, J. (1994). Thermohydraulic evaluation of two ATES projects in southern Sweden, *Proceedings of International Symposium on Aquifer Thermal Energy Storage,* pp. 73-81, Tuscaloosa, Alabama, U.S.A., November 14-15 1994

Rafferty, K. (2003). Ground water issues in geothermal heat pump systems, *Groundwater*, Vol. 41, No. 4, (July-August 2003) 408-410

Rosen, M.A. 1999. Second-law analysis of aquifer thermal energy storage systems, *Energy*, Vol. 24, No. 2, (February 1999) 167-182

Sanner[a], B. (2001). Shallow geothermal energy, *GHC Bulletin*, Vol. 22, No. 2, (June 2001) 19-25

Sanner[b], B. (2001). A different approach to shallow geothermal energy–underground thermal energy storage (UTES). *International Summer School on Direct Application of Geothermal Energy (IGD2001)*, Giessen, Germany, September 17-22 2001

Schmidt, T.; Mangold, D. & Muller-Steinhagen, H. (2003). Seasonal thermal energy storage in Germany. *Proceedings of ISES Solar World Congress 2003,* pp. 1-7, Göteborg, Sweden, June 14-19 2003

Schmidt, T. & Hellström, G. (2005). *Ground Source Cooling - Working Paper on Usable Tools and Methods*

Tenma, N.; Yasukawa, K. & Zyvoloski, G. (2003). Model study of thermal storage system by FEHM code, *Geothermics*, Vol. 32, No. 4-6, (August-December 2003) 603-607

Vinsome, P.K. & Westerveld, J. (1980). A simple method for predicting cap and base rock heat losses in thermal reservoir simulators. *Journal of Canadian Petroleum Technology*, Vol. 19, No. 3, (July-September 1980) 87-90